The
Shepherd's
Guidebook

The Shepherd's Guidebook

Margaret Bradbury

Rodale Press
Emmaus PA

Library of Congress Cataloging in Publication Data

Bradbury, Margaret, 1921–
  The shepherd's guidebook.

  Includes index.
  1. Sheep. I. Title.
SF375.B66   1977   636.3   76-30713
ISBN 0-87857-159-0

4 6 8 10 9 7 5 3 hardcover

To Brad, who encouraged
and supported my early ventures.

Acknowledgements:

To Virginia and Jim Shearer
and
Milton Johnson
and
The Shepherd magazine
for their willingness to share
their knowledge of sheep.

Contents

1

Starting
a
Flock

This is a book about sheep for the farm and homestead—a collection of information gathered from my many years of happy shepherding, from exchanging ideas with other "sheep people," and from reading about this remarkable animal, which has provided mankind with meat, wool, and leather since earliest recorded time.

EARLY SHEPHERDING

Man's best friend may be his horse, but his oldest friend must certainly be the sheep. Evidence exists to show that sheep and goats were herded and used for skins and wool since the Stone Age. The ancient Egyptians used them to tramp in newly sown grain. Primitive pastoral peoples domesticated sheep and used them for tallow, meat, milk, and wool. And by the time Abraham made his historical journeys the length of the Euphrates Valley and south into Egypt, sheep played such an important role in daily life that a man's wealth was measured by the number of rams in his flock.

Being histories of early man, the *Old* and *New Testaments* are filled with references to sheep and shepherds, in details of daily life, in song, and in story. Anyone who has raised a pet lamb will sympathize with the poor man in the book of Samuel who had a pet lamb, the only thing he owned. This lamb lived with his family and "nestled in his arms." When a rich man needed a lamb for a feast to entertain a passing traveller, he took the poor man's ewe lamb and served this to his guest.

The origin of the first sheep is confused. Some of the early ancestors were very like goats, and even today the two species are similar in many respects. Domestic sheep are believed to be descended from two wild sheep species, the Asiatic Urial and the Mouflons. Mouflons are still found in Asia Minor and the Caucasus, in Sardinia and Corsica. While the Urial lives on grassy plains in large flocks, the Mouflons are independent and daring, able to withstand harsh mountain weather conditions and leap nimbly over the roughest terrain. Mouflon-like short-tailed domestic sheep are still found in northern Europe and on the uninhabited island of Soay, northwest of Scotland. These latter sheep see man only once a year, when the people of St. Kilda come to round them up and shear them.

The Romans, well-known for their efficiency, developed sheep with fine lustrous wool from which their togas were made. Roman citizens took great pride in their flocks, which, through the very best care, produced the finest wool in the world. When the Romans invaded Britain they took with them some of their flocks and greatly improved the rather scruffy native sheep they found. During their occupation of the country they encouraged the manufacture of quality woolen materials for export to Rome. After the collapse of the Roman Empire, little headway was made in developing the wool business in England until the twelfth century.

Up until that time, sheep were used for their wool and skins and for the value of their manure in maintaining the fertility of the soil. Then, when living conditions became more settled, the wool industry flourished, bringing trade and wealth to the country. The manufacture of woolen goods became so important that for many years the export of raw wool was prohibited. Flemish weavers were encouraged to settle in England and brought with them their skills in weaving, dyeing, and cloth making. By the time of Elizabeth I, wool was the country's chief source of revenue. Parliament recognized the extent of the prosperity that wool brought to the land and decreed that the Lord Chancellor and the judges should sit on woolsacks in the House. At first, these gentlemen sat on real woolsacks, and now they sit on ottomans stuffed with wool that bear the same name.

During an embargo on the export of raw wool. Parliament decreed that all dead must be buried in a woolen shroud (the customary burial garment), and that a notation that this had been done must be made by the parish priest. This edict gave rise to a parish register called the *Wool Book,* and the custom lasted for some thirty years. In some instances it was recorded that a "flannel" garment was used, the word flannel being a corruption of the Welsh word for wool, *gwlanen.*

It is interesting to note that, although discoveries have been made of various shepherd's tools and possessions (such as crooks and shears), very few shepherd's smocks have been found, leading to the supposition that shepherds preferred to be buried in their linenlike twill smocks. It was usual to put a tuft of wool in the shepherd's hand before the coffin was closed, so that when the departed soul arrived at the gates of heaven, Saint Peter would know that the newcomer was a shepherd and would therefore excuse the man if he had not always attended church on Sunday.

With the eighteenth century and the industrial age came a rapidly expanding population. The need for more food, especially meat, and the introduction of improved methods of farming—with systems of crop rotation and provision of winter forage for sheep—made possible the development of a meatier animal. Many of the new breeds that were developed during these years have been imported into the United States and are the foundation for the breeds now found here. Some flocks exist in their original breeds, such as the Down breeds, the Cheviots, Suffolks, Romneys and the long-wool breeds (Lincoln, Leicester, and Cotswold).

From these, new American breeds such as Montadale, Columbia, and Targhee have been developed.

While the American bighorn sheep is the only sheep native to the United States, the ancestors of our domestic sheep were the Merinos brought here by Francisco Coronado, governor of New Galicia, Mexico, when he was sent by the Spanish Viceroy in Mexico to invade the "lands to the north." In spite of setbacks, uncertainties, and deprivation, the flocks of sheep gradually increased from the eighteen survivors of Coronado's original flock to more than fifteen-thousand head by 1780. Some of the animals were moved to the west, where the missions were noted for their well-cared-for sheep, and others went north to New York and Ohio.

The first British sheep imported into this country were brought into Virginia early in the 1600s, soon to be followed by others coming into Massachusetts. Later in the century the export of ewes or lambs from England to America was forbidden, so strict measures were taken by the owners of the sheep already in this country to protect them and ensure their proper care. The town shepherds were important persons, with the responsibility of finding pasture for their flocks, for getting them shorn, and for protecting them from thieves or dogs. Dogs that chased or killed sheep were liable to be hanged. In Connecticut, every person was required to work for one day a year clearing land for pasture. Gradually, by the use of imported long-wool rams and through rigidly selective breeding, great improvements in weight and quality of fleeces were achieved. At the same time the increased importance of the lamb and mutton trade brought about the development of a dual-purpose animal.

RAISING SHEEP TODAY

As a farm animal, the sheep is unsurpassed in its adaptability. Today there are thousands of farms and homesteads in the United States with areas of five to five hundred acres where farm flocks thrive and do a great deal of good in keeping down brush and weeds and in fertilizing the soil, while at the same time producing high quality meat and wool. With good pasture, adequate shelter, and a minimum of care, sheep will provide their owner with a return for his investment equal to that received from cattle or hogs. Their temperament and relatively small size make them easy for women and older children to handle without fear of injury. Working with them is neither hard nor time-consuming, and the shepherd who learns to understand sheep and who enjoys caring for them will succeed, while the one who treats them only as scavengers and weed killers will get a disappointing return.

Sheep graze and do well both on land that is unfit for or inaccessible to plowing and on grasses not suitable for grazing other animals. While they are grazing they are converting this grass, which would otherwise go to waste, into protein in the form of meat and wool. They are careful eaters, choosing a variety of plants, herbs, and grasses, and at the same time keeping down such pesky greenery as crabgrass, purslane, dandelions, and poison ivy; in early spring their appetite for wild onions is very useful, as the milk of dairy animals takes on the flavor of garlic when the cows are first let out on newly growing spring pasture.

Because they are ruminants, sheep need very little grain, requiring this type of feed only at lambing time in areas of the country where they must be fed over the winter, and in late spring when lambs are given some grain to give them a finish before marketing.

Sheep will provide income from the sale of meat, wool, live animals, and from sheepskins. Their products can be prepared for use in the home with very little difficulty, and recipes, instructions for home butchering, and methods of tanning are given later in this book. Wool from the fleeces can be used in a variety of rewarding ways: a well-cared-for fleece can be spun into yarn for knitting or weaving, and the comforters made from wool batting are wonderfully warm.

A few factors on the debit side must be considered. Sheep are defenseless animals, and their tendency to scatter in panic makes them easy prey to dogs and coyotes. Even if not killed by dogs, they may die of shock and exhaustion after being chased and worried by them. Stray dogs may cause trouble, but local pets present a greater danger. Running a goat with the flock will afford some protection because goats are less afraid of dogs, and the sheep will rally round them.

As parasites, internal and external, can be a problem, you must be prepared to follow a regular program of pasture rotation, worming, and external parasite control if the flock is to remain healthy and productive. This is a time when an ounce of prevention is worth about five pounds of cure.

Because of the difficulties of complying with so many federal, state, and local regulations, many custom butchers have closed their businesses. The best way to find someone to slaughter lambs is to ask a neighbor who also has livestock. Or, if space and equipment are available, home butchering is possible. Finding a market for lambs will not be difficult; as soon as the neighbors know you have lambs for sale, they will line up at the door.

In some suburbanized areas, not all veterinarians are willing to work with sheep, and it is important to find one who understands sheep and is interested in not only treating sick sheep but in helping maintain the good condition of the flock.

No class of livestock production is without its risks and hazards, but these can be overcome with a little care and forethought. The successful shepherd gets to know the animals by watching them and studying their natural dispositions. Sheep don't want to be fussed over. They need some extra care at lambing time and shearing, but for the most part, if you will do your part, as needed, the sheep will do the rest.

BREEDS AND TYPES OF SHEEP

What do you want from your sheep? Are they to provide meat and wool? Do you want to sell lambs and Easter lambs? Or do you want to raise purebred breeding stock? Many different factors influence the choice of breed—locality, climate, type of terrain, and intended use should all be considered before a decision is made.

SELECTING GOOD STOCK

Brief descriptions of some of the breeds more commonly found in this country are given below, together with the names of their breed registry associations from whom information and descriptive material is available. Whether the flock is to consist of purebred or crossbred ewes, or even all sorts, the ram should be a purebred, selected from a breed which will endow his progeny with the qualities you are seeking. A mixed-breed ram is no economy; he may turn out to be just great, but more often than not he will sire a hodgepodge of lambs of all shapes and sizes. The characteristics of the ewes will be inherited by their own offspring, and those of the ram by the offspring of every ewe in the flock, so it is important to find a ram that is a good specimen of his breed, active and fresh, well-built and with good body conformation, not too fat, and of the type you want to develop in your flock.

It is not necessary to buy purebred ewes in order to have a good flock. Crossbred (or grade) ewes, as long as they are healthy and well-built, can be useful and profitable, and the investment will be much lower. If you're buying a whole flock, look for one in which all the sheep are similar in size and appearance; or, if buying individual sheep from different places, try to follow the same principal of selecting a uniform type. Whether purebred or mixed, the ewes should look alert, skin should be pink and smooth, the gums and mucous membranes of the nose and eyelids should be bright red, and the udder firm and pliable, with no lumps and, of course, two teats. There should be no diarrhea or runny noses. Check for sore feet: sheep, like people, cannot be happy when their feet hurt.

To inspect the wool and skin, place the hands with the palms together on the sheep's flank and slide them into the wool, then lay the palms upward. The wool should be dense and bright, clean, and free from parasites. Never open the fleece on the sheep's back, as this will let rain and dust onto the skin.

Teeth are a reliable indicator of the sheep's age, up to about five years. After this time the teeth become progressively more worn and some break off, a condition called broken mouth. Because a sheep has a broken mouth is no reason not to buy her, as long as she is otherwise well-fleshed and healthy looking; many ewes of six or seven have several more productive years ahead of them and can provide the foundation for a good flock.

If the prospective purchases are in full fleece, try to imagine what they will look like when shorn. Place a hand on the back, just in front of the haunches; the body should feel wide and firm, with no ridge or backbone sticking up. The ribs should be firm and well set, and the chest wide and deep. Avoid sheep that are fine-boned, skinny in frame, or have overly long legs. Of course, appearances aren't always everything. A ewe who has recently reared a pair of fast growing lambs will not have the full bloom of a young ewe who has had no demands made on her, and a beautiful, smoothly rounded ewe with elegant wool may look that way because she is barren and living a life of well-fed ease.

Lamb's mouth with 8 incisors. Temporary teeth called milk teeth.

Yearling's mouth with 1 pair of permanent incisors.

Two-year-old's mouth with 2 pairs of permanent incisors.

Three-year-old's mouth with 3 pairs of permanent incisors.

Four-year-old's mouth with 4 pairs of permanent incisors.

Broken mouth, which may occur at about 6 years of age. A sheep that has lost all the incisors is called a gummer.

Unless there is a good market for breeding stock, it is of no special advantage to buy purebred ewes. The selling of breeding stock requires an extensive knowledge of sheep breeding, a good business sense, and meticulous record keeping. Good grade ewes will produce desirable wool and lambs, and the initial outlay will be smaller. Some mixed ewes are, of course, more mixed than others. Perhaps there are not many sheep in your locality, but you know of a flock for sale that has a variety of different shapes and sizes. If these are healthy and active, there is no reason why a good flock could not be built up by using a purebred ram and keeping only the best of his progeny for further breeding. The ram lambs, of course, should not be used for breeding. The important thing is not the breed, but the production of marketable animals of good meat and/or wool type. One word of caution—odd small flocks of ewes that are for sale are frequently accompanied by an equally odd ram. Don't try to keep him if you want to try to use a purebred ram to improve the flock because he will very likely find his way over the fence or under the wire to the ewes—and there goes your planned breeding.

Before deciding on any one breed or type of sheep, you would do well to visit other flocks in the area to see what kind of sheep are commonly kept, what kind of market there is, and how well such sheep would fit into your plan. The county agent is another good source of information, and so is a local veterinarian. There is no best breed to buy, as the best breed or type is the one that will be most suited to the environment in which you plan to raise it. It is well to start with a few sheep the first year—ten is a good number—and to try to buy them in summer so that you have the rest of the summer and fall to get to know them before the critical period of lambing time. Always start with at least two sheep, for they suffer from fear and loneliness when by themselves, and need the company of their own kind. You will very likely have all sorts of questions during these first few months, so visit neighbors who have sheep, or go to local fairs. Most sheepmen love to talk about their animals and will go out of their way to be helpful.

PROFILES OF THE BREEDS

There are some two hundred breeds of sheep, more than in any other class of animals, and of these about twenty-five are found in the United States. Remember, no one breed is the best; the best breed is the one most suited to the conditions of climate, terrain, and management intended for it.

MERINO

The Merinos were the ancestors of our domestic sheep, brought here from Spain via Mexico. These early sheep were noted for their fine and beautiful wool, and their descendants, the Merinos, Rambouillets, and Debouillets of today, are fine-wooled sheep, hardy, gregarious, and long-lived. Merinos are well suited to range conditions and, like

Merino.

the Dorset and Tunis, will breed out of season. This breed has contributed more to the development of the present-day sheep industry than any other breed. It has a deep-rooted flocking instinct that makes it possible for one herder to watch over a large flock, and even a strain of Merino in a sheep of another breed will make that sheep more likely to resist an attack by a predator. The early type Merinos, with their deeply folded skin, have largely given way to the modern Delaine Merinos, developed in Ohio, West Virginia, and Pennsylvania, which have better meat qualities and a skin free from folds.

There are not very many established Merino flocks from which to buy stock, but information may be obtained from A. L. Liming, Lou-Ida Farms, Mineral Ridge, Ohio 44440.

Rambouillet.

RAMBOUILLET

The Rambouillet, named for a town in France but largely developed in Germany and the United States, has been bred as a dual-purpose sheep. The skin is almost free from folds and the wool is long staple, of fair density and uniform shrinkage, making it an ideal fleece for hand spinners. The breed is very hardy, can survive rough conditions and sparse forage, and has a strong flocking instinct. Contact the American Rambouillet Sheep Breeders' Association, 2709 Sherwood Way, San Angelo, Texas 76901.

DEBOUILLET

The name signifies the origin of the breed—"De" from Delaine and "bouillet" from Rambouillet. Similar in appearance to the Rambouillet, these sheep have many of the same characteristics and carry a fleece of long-staple, dense, fine wool. Write Mrs. A. D. Jones, Debouillet Sheep Breeders' Association, Roswell, New Mexico 88201.

The breeds more commonly found in the East are the medium-wood breeds—that is, the wool is medium in length, weight, and fineness. The Down breeds are medium-wooled and come from the South of England where the rolling hills are called downs, and were bred primarily for mutton. They

11

are popular for farm flocks, and the rams have been used extensively on western range ewes.

The five Down breeds are Southdown, Shropshire, Oxford, Hampshire, and Suffolk.

SOUTHDOWN

This is the real mutton-type breed and is known in nearly all of the sheep raising countries in the world. The best of these sheep native to the South of England were selected for importation into this country and are the foundation of all the other Down breeds. The Southdown is the smallest of the mutton breeds, having moderately short legs and a chunky body; the ewes weigh about 125 pounds and the rams reach about 275 pounds. The fleece is short, dense, and fine, yielding eight to ten pounds of wool per clip. These sheep are ideal for the farm flock because of their smaller size, meaning that a larger number can be grazed on a given area, and because of cheaper maintenance and early maturing lambs, which are ideal for the hothouse lamb trade. Hothouse lambs are young milk-fed lambs specially reared for Jewish Passover and the Greek and Italian Easter market. The carcase value cannot be excelled, because of the lean meat-to-bone ratio. Southdowns are quiet and easily handled. The

Southdown.

rams are much in demand as flock sires, and Southdown-cross lambs will reach market weight on grass alone. For information write Howard's Southdowns, Mulhall, Oklahoma 73603.

SHROPSHIRE

The Shropshire is a good middle-of-the-road sheep. It is a little larger than the Southdown, and the body is thick and well fleshed, carrying a fleece of fairly long, medium-fine wool. Clips of ten pounds of wool per shearing can be expected. While they rank lower than the Southdown in early market readiness, they are still an excellent all-round animal for the farm flock. Write American Shropshire Registry Association, Inc., Mrs. E. R. Glasgow, P.O. Box 1970, Monticello, Illinois 61856.

Oxford.

OXFORD

The Oxford is well suited to the farm flock and rates the largest of the Down breeds. Rams will attain a weight of as much as 350 pounds and the ewes 250 pounds, but because of their relatively docile nature, Oxfords are easy to handle (despite the handler's expression), regardless of their size. They will shear ten to twelve pounds of wool of relatively long, loose fleece. The ewes are prolific and good mothers. The lambs are large and vigorous at birth and make satisfactory growth. They are well suited to the farm flock. Write

13

Barton's Oaklawn Farm, 920 Miller Road, Plainwell, Michigan 49080; or American Oxford Down Record Association, P.O. Box 401, Burlington, Wisconsin 53105.

HAMPSHIRE

In western range flocks, sires of this breed are probably the most widely used for crossing with range ewes. They are an excellent meat type, renowned for the rapid growth rate of the lambs. The ewes are prolific and good milkers, and are of a medium size and quiet disposition that makes them desirable for a farm flock. The fleece shears an average weight of seven to eight pounds of medium-quality wool. Write the American Hampshire Sheep Association, Route 10, Box 199, Columbia, Missouri 65201.

Hampshire.

SUFFOLK

The Suffolk is a fine, handsome sheep of good size that has shown a great increase in popularity during the last decade. An excellent sheep for the farm flock, the ewes are good mothers, producing one or two fast-growing lambs. Suffolks experience little trouble when lambing because of the small heads and shoulders of their young. They are a medium-weight animal, with face and legs free from wool, and are excellent foragers. Their greatest deficiency is the light, short fleece, which makes them unsuitable for those who want to produce wool for spinning and weaving. Write the National Suffolk Sheep Association, P.O. Box 324N, Columbia, Missouri 65201.

Suffolk.

Cheviot.

CHEVIOT

The Cheviot is a beautiful sheep, with a smooth white face free of wool, short legs, and a blocky appearance. Cheviots originated in the Cheviot Hills of the border country between England and Scotland, where the climate is harsh and the conditions rugged. They are agile and exceptionally hardy, and their lambs can survive being born outside under

15

poor weather conditions. The ewes are good milkers and attentive mothers. The Cheviot produces good meat lambs, which are usually born later than those of the Down breeds. They are slightly slower growing, but of good quality. The fleece is light-shrinking, and the wool off a well-kept Cheviot will provide a beautiful material for hand spinners.

This is a good breed for a farm having a lot of rugged hilly pasture, as the sheep are first-class foragers, surefooted, and independent. Cheviots are reputed to be flighty and hard to manage, but if properly and quietly handled they are not more nervous than any other breed. Cheviot rams are frequently used on first-lamb ewes and ewes of the chunkier Down breeds because their lambs have small heads and cause less difficulty in lambing. For more information, contact the American Cheviot Sheep Society, Inc., Dr. Larry E. Davis, Rural Route 1, Carlisle, Iowa 50047.

Dorset.

DORSET AND DORSET HORN

These two breeds are identical, except for the absence of horns in the polled strain. Probably their best known characteristic is that of producing two lamb crops each year. The ewes are good milkers and attentive mothers. Originating in the hilly country in the Southwest of England, the Dorset is its native land. In type, it compares favorably with other medium-wool breeds and is much in demand in the eastern United States, where there is a strong market for winter, or hothouse, lambs. The breed will do well in the farm flock, but it should be noted that the ewes must receive proper nourishment and care if they are to be expected to raise two families per year. For many years the breed was never considered suitable for the range country and never crossed the Mississippi River, but today they are popular throughout the West, clear to the Pacific Coast.

History tells us that during the time when Spain was trying to conquer England, Merino sheep were likely brought into the Southwest of England and crossed with the horned sheep of Wales. Although the sheep which resulted from that cross are different today because of changes in environment and breeding, some characteristics are still found, proving that the Dorset and the Merino carry similar bloodlines. Write the Continental Dorset Club, Marion A. Meno, Secretary, P.O. Box 577, Hudson, Iowa 50643.

Columbia.

COLUMBIA

The Columbia was the first breed of sheep developed in this country, in honor of which the breed bears its name. It is a crossbreed, descended from a long wool and fine wool foundation, producing a medium-fine wool and at the same time a sheep well suited to range conditions. It is a utilitarian breed. The lambs are of good market quality and the wool is excellent, the latter being inherited from its Rambouillet ancestry. Such sheep adapt well to farm flock conditions and are increasing in popularity in the East. Write Columbia Sheep Breeders Association of America, Richard L. Gerber, P.O. Box 272E, Upper Sandusky, Ohio 43351.

CORRIEDALE

This is the oldest of all the crossbred wool breeds, originating in the late nineteenth century in New Zealand, where both meat and wool production are sought. From its long-wool ancestors, Lincoln and Leicester, and from Merino ewes, the Corriedale has inherited good meat conformation and a dense fleece of very good quality.

Corriedales are smaller than Columbias and less rugged, but this would not be a disadvantage in the farm flock, where they are an economical breed by virtue of their beautiful fleeces of long-staple wool, with its distinct crimp, plus their general production ability in giving more pounds of lamb and wool per pound of body weight than other range breeds. Write American Corriedale Association, Inc., Russell Jackson, Secretary, Box 29C, Seneca, Illinois 61360.

FINNISH LANDRACE (FINNSHEEP)

This breed is native to Finland and is comparatively new to this country. It has gained popularity rapidly during the last few years and is known chiefly for its ability to produce litters of lambs and for its early maturing quality. The Finnsheep ewe is an excellent mother and a heavy milker, easily nursing three lambs under farm flock conditions. Research indicates that these characteristics are readily transmitted to cross-

Corriedale.

bred ewes with Finnsheep breeding. The multiple-lambing characteristic requires careful management to ensure that all the "bonus" lambs are raised. Write Finnsheep Breeders Association, Inc., Rural Route 3, Pipestone, Minnesota 56164.

MONTADALE

The Montadale was developed during the 1930s by E. H. Mattingly of St. Louis, Missouri, using first a Columbia ram on Cheviot ewes, and then reversing the process. The result is a good meat-type animal with the head and legs free of wool and with the stylish appearance and agile body of the Cheviot. A wool clip of ten to twelve pounds may be expected. More information is available from Montadale Sheep Breeders Association, Miss Mildred Brown, Secretary, P.O. Box 44300, Indianapolis, Indiana 46244.

TARGHEE

This is the second new breed developed in this country. At the United States Sheep Experiment Station in Idaho, ewes of Corriedale x Lincoln-Rambouillet breeding were mated to outstanding Rambouillet rams, and the offspring of these matings were inbred to produce the Targhee. The breed gets its name from the Targhee National Forest, in which the flock graze during the summer. They were developed under range conditions and selected for production performance. It is a medium-size, white-faced sheep, carrying a fleece of about eleven pounds. The ewes are good milkers and adapt well to farm flock conditions. Write U.S. Targhee Sheep Association, Box 2513, Billings, Montana 59103.

TUNIS

This is a medium-size breed with a reddish or tan face and legs, and low pendulous ears. The ewes are hardy and good mothers and, like the Dorset, they will breed out of season. The wool is white, of medium length, and of good quality. Write National Tunis Sheep Registry, Nesbitt Road, Attica, New York 14011.

The long-wool breeds are Cotswold, Leicester, Lincoln, Romney, and Karakul. Sheep of the first three long-wool breeds are not plentiful in the United States, and information concerning them should be sought from state departments of agriculture.

LEICESTER

The Leicester is the ancestor of most modern long-wool sheep. George Washington used Leicester sheep in improving his flock at Mount Vernon, but there are very few Leicesters of either the English or Border type in the United States today.

LINCOLN

The Lincoln originated on the east coast of England, in Lincolnshire, and is still the principal long-wool sheep in the United States. It is a very old breed, large, coarse, and slow-maturing with a heavy fleece. Lincolns are reputed to be the heaviest sheep in the world. The breed was improved in the nineteenth century by the introduction of Leicester blood,

Romney.

and sheep of the improved strain were imported into this country in the mid-1800s.

COTSWOLD

This is a very large sheep with coarse wool that hangs in ringlets eight to fourteen inches long, all over the body.

ROMNEY

These sheep originated in the Romney marshes of Kent, in England, where there is abundant forage, but where the wind and rain howl across the landscape. Romneys are adaptable to many different environmental conditions. They are said to be less susceptible to foot rot than other breeds, and are hardy and of good size. A yearly clip of ten to twelve pounds of quarter-blood wool can be expected. Romney wool is in great demand by handweavers and spinners because it is a bit larger in diameter and has a staple length of from four to seven inches. Contact the American Romney Breeders Association, Dr. John H. Landers, Jr., Secretary, 212 Withycombe Hall, Corvallis, Oregon 97331.

KARAKUL

The Karakul is a fur sheep, with coarse and stringy wool and progeny of poor meat-type. The skins of baby lambs, with their tightly curled wool, are used in the fur trade and are known as Persian lamb. There is little demand in this country for such skins, but the hides of mature sheep are popular for use as sheepskin rugs, and a Karakul ewe bred to a Suffolk or Hampshire ram, or vice versa, will usually produce a black lamb and, eventually, an attractive black or brown pelt. The wool of Karakul ewes and Karakul crosses is not solid black, but is a fleece of black, brown, and gray shadings that can be spun into attractively patterned garments. Broadtail, the most valuable pelt, is produced from prematurely born or stillborn lambs. Such deaths are the result of accidents or abortions and are not forced.

21

SCOTTISH BLACK-FACE
(BLACK-FACED HIGHLAND)

These sheep, when bred to a Southdown or Shropshire ewe, will produce an excellent milk lamb. The ewes of these breeds are hardy and good mothers. The fleece consists of a long, coarse outer coat and a very fine inner coat. Black-face Highland and Karakul are the only carpet-wool (long coarse wool) breeds known to any degree in this country. The importation of any breed from another country, such as the Black Welsh Mountain sheep for black wool, is a long-drawn-out process, involving lengthy quarantine in Canada and endless red tape. For spinning and weaving enthusiasts, *The Shepherd* magazine carries a registry of black sheep breeders.

PET LAMBS

> *Mary had a little lamb,*
> *Its fleece was white as snow,*
> *And everywhere that Mary went,*
> *The lamb was sure to go.*

There really was a Mary and she did, indeed, have a little lamb. The Mary in the poem, which was printed in McGuffey's *Second Reader,* was Mary Sawyer, of Sterling, Massachusetts, and her little lamb was one of a pair of twin lambs born on her father's farm. The ewe had neglected one of her lambs and Mary found it nearly dead with cold, but with the little girl's care and plenty of warmth, the lamb survived the night, and by morning was able to stand up. From then on, the lamb throve and became Mary's special pet.

Lambs make delightful pets, especially for children; they are gentle and affectionate, don't bite, and don't require any unusual feeding. They should, however, be kept outdoors— the basement or garage is no place for them. It is best to have two lambs, as one alone will be lonely and long for the company of its own kind, although they are known to do quite well alone if they have plenty of attention from humans.

Ideally, pet lambs should have a small grass enclosure with a comfortable shelter in a protected corner, where they can be warm in winter and shaded from the hot sun in summer. For two lambs, a shed about five feet wide, four deep, and four high should be adequate. In one corner a bracket could hold a water pail, and a small sloping rack could be used for hay and grain. The rack can be made from 1x2s, or a metal combination hay and grain rack for one or two sheep can be purchased from NASCO livestock supply, Fort Atkinson, Wisconsin 53538. To make a bracket to keep the water pail from tipping over, cut out a piece of thin metal one inch wide and long enough to fit around the pail with a two-inch overlap. Join the two ends to make a ring, drill a hole through each end, and drive a bolt through the wall of the shed and through the holes in the metal. Fasten with a lock washer and nut. The same method can be used for the salt container. The roof of the shed should slope, and if there are no trees for shade, a two-foot overhang will give extra shade and protection from flies in the summer. Such a shelter could be built with 4x4s set in the four corners for support and sunk in the ground, or the whole thing could be set on skids, like range shelters for chickens, so that it could be moved to a different location when the surrounding ground became trodden down.

Sheep realize where they live and, if they are given enough attention, soon become attached to the people who care for them, so that it is usually possible to let them out of their enclosure to graze other ground without the risk of their running away. If this cannot be safely done, a second enclosure will be needed so that the ground can be rested and to prevent parasite build-up. If the sheep are to be let out to roam freely, remember that they enjoy the taste of other greenery than grass and will relish newly growing lettuce, tomato plants, tulips, and many other garden plants—a liking which will not endear them to the gardeners in the family; in fact, it sometimes seems that the harder the plants are to grow, the more the lambs like them.

Ewe lambs should always be chosen if the pets are to be kept more than one season. Ram lambs are nice when they are really small but they grow up to be more aggressive and less affectionate, on the whole, than ewes. This is common to all livestock. Especially if there are females in heat in the vicinity, the males can become bad-tempered and even dangerous. Wether lambs (castrated males) are less cantankerous, but they too can be a problem when they reach maturity because of their greater size and the fact that, even if they are only playing, it is no joke to be butted by two hundred pounds of bumbling ram who remembers playing with you when he was a little lamb. The ewes seem to become progressively more matronly and settled, and will seldom do anything more aggressive than push their way to the feed bucket.

Feeding pet lambs need be no great task. First of all, remember that they do not know when to stop and must therefore be given only as much grain at one time as they can safely digest—about ½ pound when young, increasing to one pound per day in winter for a full-grown sheep. Hay, which is roughage, is different; they may be fed hay free-choice, as long as it is not alfalfa, and can be given a week's supply at one time. (Alfalfa hay is rich and may cause bloat if overeaten.) A V-shaped hay rack is best; it will keep the hay seeds from getting in the wool round the neck and spoiling the fleece. Clean water and a salt-plus-mineral ration should be available at all times.

If the lambs are very young when purchased, they will need to be fed milk from a bottle until they are well started on hay or grass and grain. An ordinary baby nursing bottle with the hole in the nipple slightly enlarged is satisfactory. Feedings should consist of six to eight ounces of whole milk every four hours, omitting the middle-of-the-night feeding (see chapter 4, *Orphan lambs*). When they have settled down to eating hay and grain, the milk ration can be cut down gradually by giving less at each feeding. At three weeks, a small pan of grain can be offered, and the lambs will nibble a little; put out only a small quantity because they will waste most of it at first. As soon as they discover that the grain is good to eat, give two grain feedings, one morning and one evening, consisting of a double handful of oats, or oats and cracked corn. As they get used to this, the feed can be changed to any good commercial horse and pony feed, but not before they are at least six weeks old because there is a danger of their choking on the kernels of whole corn that are used in horse feed. As they eat more hay or grass and grain and start drinking water, decrease the milk feedings slowly. Reduce the quantity of milk, but don't add water to the bottle, because this will mean that the lamb has to take in too much liquid to satisfy its hunger and will have no room in its stomach for the solid foods that are required in order to start the four stomachs working. When the grass is growing well in the summer, the lambs will need no other feed, although a handful of grain once a day keeps them coming to one spot for this feeding and thereby makes it possible to watch them for signs of lameness, worms, or minor injuries. They will also be more tame if they are used to being hand-fed.

When winter comes, the lambs will be almost full-grown, and in areas where grass is not available over the winter, a ewe will eat about three pounds of hay per day, plus a little grain. Cold is usually no problem, since sheep have four to six inches of insulating wool on their backs, but rain, particularly freezing rain, can give them colds and pneumonia. As long as the shed is kept open to give them free access, the sheep will go inside when the weather is harsh.

Ewe lambs that are well grown out by November can be successfully bred during their first winter or, if they are not too well developed, they may be bred equally well in their second year. It is not practical to keep a ram for one or two ewes; a better arrangement is to find another sheep owner who will either lend a ram for a few weeks or who will keep the pet lambs in his own flock for a few weeks until they are bred. In either case, the ewes and ram should be checked for worms and parasites before leaving home and again before returning. Young maiden ewes come into heat later in the year than ewes that have already borne lambs and should not be put with the ram until at least November. It would, of course, seem easier just to buy a ram. But this would mean that if the ewes were to have one or more ewe lambs, these would have to be sold, or the ram would have to be kept separate from them the following breeding season, to avoid inbreeding. Then, too, it would hardly pay to buy a good purebred ram to breed two or three ewes—the tempta-

25

tion would be to use any little old ram lamb that happened to be available.

Finding a shearer for the lambs in the spring may not be easy. The problem may be solved by taking the sheep to another farm where sheep are being sheared; if this course is followed, make sure the lambs do not "mingle" and thereby expose themselves to internal or external parasites, or even, perish the thought, carry in such parasites. When they return home, take a handful of Marlate, or a pint of Diazinon sprinkle, and run this over their backs as a precaution against ticks or lice. In a few days, have them checked for worms and look at their feet to make sure the hooves are not overgrown. The value of the wool varies widely from year to year, but it is always worth something and can be sold at the local county wool pool; or, the shearer may accept the fleece in payment for shearing.

Pet lambs love the attention they get from making public appearances. Lambs from our flock have taken part in Christmas Nativity tableaux, they have been to school for nature study days and to church for Shepherd's Sunday, and have served as models for 4-H instruction, among other things. The easiest way to show them, we found, was to borrow a child's playpen and cover the floor with plenty of newspaper. With a little hay and water to keep them contented, the lambs would be quite happy while the children gathered round to pet them, without danger of overenthusiastic handling or of the lambs running away.

Lambs that have been pets adapt quite easily to flock life, either as additions to existing flocks or as the foundation for a new flock. Our first pet lamb, Phoebe, had a gentleman caller for the month of November, and the following spring she produced a lamb of her own; the next year we had three lambs, and the following year the shed had to be enlarged to accommodate the growing flock.

2

Housing, Equipment, and Fencing

Most properties large enough to support a flock will have some sort of barn or shed that can be adapted to house the sheep. In areas warm enough to be free of frost and snow in winter, no shelter is needed, except in summer when shade must be provided because of flies. Sheep are not fussy, and will do nicely in a barn or shed having a clean, well-bedded floor and racks to keep their feed off the ground.

USING EXISTING BUILDINGS

The big barns found on farms will adapt very well for sheep with a few alterations. First of all, the stalls and mangers should be thoroughly cleaned out and aired, broken windows should be repaired, and the accumulation of old buckets, wagon trees, one-pronged pitchforks, and cans bearing unknown contents should be thrown out. Once the building is clean, make a plan of the floor space needed, allowing for a few stalls to be used as lambing pens, a pen for the ram, and an area for a lamb creep.

Water is heavy, and sheep drink several gallons a day during their lactation or when being fed dry rations—sometimes even more when they know you are carrying it for them. So, it is a great saving of time and effort to have water piped into the barn and, if possible, a means of heating it.

Electricity can make it possible to have adequate lighting, as well as warming tapes to prevent water pipes freezing, immersion heaters for water troughs, and heat lamps for lambs that may be chilled. If the barn has been unused for some time, it is well worth having an electrician check the wiring, since too many old buildings have a Rube Goldberg collection of frayed wires and rusted sockets. Kerosene lanterns provide a nostalgic touch and are best left hanging unused on the wall as a reminder of simpler days. They are dangerous and have no place in today's barns.

Doors should be wide enough for the sheep to move in and out without crowding. Because damp, not cold, is the chief problem in winter, it is neither necessary nor desirable to have a tightly closed building. One side of the building could be opened up to make the largest possible loafing area. This should be on the sheltered (probably south) side of the building so that maximum sunshine reaches the interior. Screens can be hung in summer to darken the barn if flies are bothering the sheep. All steps and sills should be removed or made shallow enough that ewes heavy with lamb will not have to jump in and out. The sheep can spend their time here until their lambs are born, and then mother and young can be moved into one of the stalls for three or four days. If enough room is available, another open area can be arranged to serve as a nursery barn. The layout of the barn will depend on the shape and condition of the existing building, the main considerations being that it is light and airy, with convenient access to feed storage and water supply, and with a well-drained floor and yard.

These alterations should be made before the sheep are installed. If the animals are allowed into a barn that is full of old bedding and debris, there is a danger of their eating something poisonous or injuring themselves on a projecting nail or piece of wood. The shepherd also runs a risk of injury.

NEW HOUSING

When new housing is to be erected, the buildings need not be elaborate or expensive. A simple pole barn of a single story, with hay and grain storage area on one side, will do very well. Such a building can have many uses when the sheep are not occupying it. The barn should be three-sided, open on the south side, and wired for electric light. Floors may be packed earth, gravel, or concrete. Allow twelve square feet per sheep, and provide for at least ten-foot headroom. Where the winters are cold, one end of the barn can be walled off to provide a sheltered lambing area, lamb pens, and a hospital pen. A ventilation fan will be desirable in the enclosed section. The Midwest Plan Service of the North Central Land Grant Universities (Ames, Iowa 50011) has compiled *Sheep Housing and Equipment Handbook,* (MWPS-3), containing excellent plans and directions for barns, together with information on manure disposal, ventilation systems, and feed handling. In Pennsylvania, the Agricultural Engineering Extension (204 Agricultural Building, University

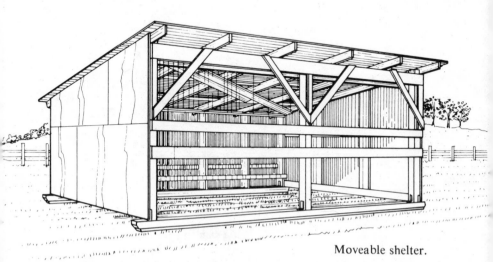

Moveable shelter.

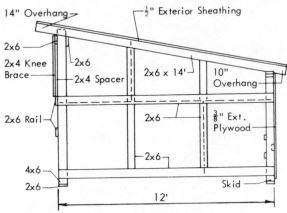

End Framing

Note:
Plywood sheets indicated by dotted lines. Cover roof with roll roofing.

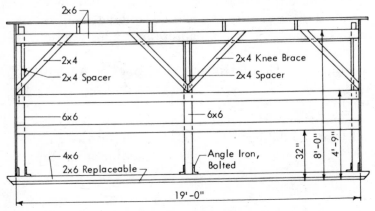

Front Side Framing

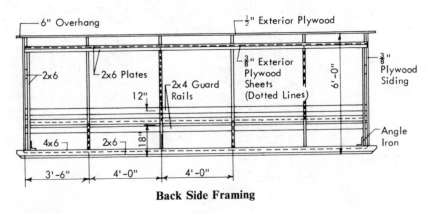

Back Side Framing

(Reprinted with permission from MWPS-3, *Sheep Housing and Equipment Handbook*, Midwest Plan Service, Ames, Iowa 50011.)

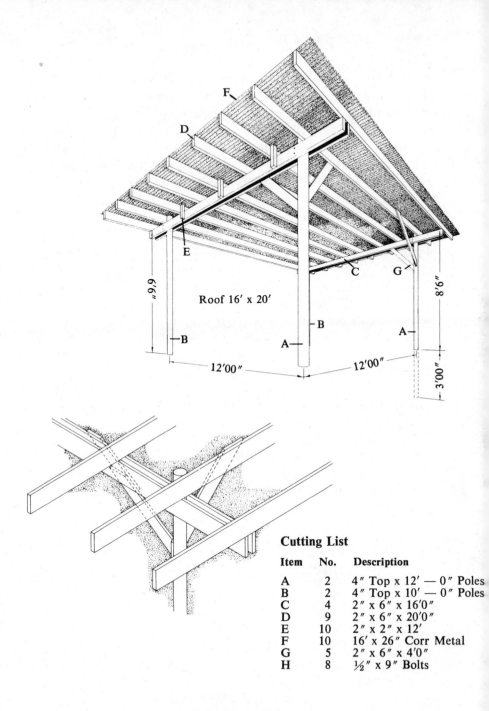

F

D

E

Roof 16' x 20'

6'9"

C

G

8'6"

B

B

A

B

A

A

3'00"

12'00"

12'00"

Cutting List

Item	No.	Description
A	2	4" Top x 12' — 0" Poles
B	2	4" Top x 10' — 0" Poles
C	4	2" x 6" x 16'0"
D	9	2" x 6" x 20'0"
E	10	2" x 2" x 12'
F	10	16' x 26" Corr Metal
G	5	2" x 6" x 4'0"
H	8	½" x 9" Bolts

Stationary shade (adapted from Iowa State
University sheep equipment plans).

Park, Pennsylvania 16802) has plans for farm buildings, and the Northeastern Regional Agricultural Engineering Service has issued plans and manuals in *Build a Pole Barn* (NRAES, Filey Robb Hall, Cornell University, Ithaca, New York 14850).

Because prices change so rapidly, it would not be practical to estimate here the costs involved in construction. With the rise in building costs there is an increased need for very careful planning, so that the best possible use may be made of other people's experience with design and construction.

Sheep and their young, like other farm animals, need to be warm and dry. Proper feeding will keep them warm, and proper ventilation and bedding will keep them dry. A barn or shed should never be closed: a wide open door is much better than a lot of small open doors and windows because there will be less draft. Sheep generate body heat that will warm the barn area, just as cows do, if the ceilings are not too high. As the warm air is exhausted from the barn area it will pick up moisture, keeping the inside drier and healthier. In winter, insulation can be provided by stacking bales of straw along the coldest walls.

A loafing pen around the barn will prove very useful. The sheep can be held there for dipping and worming and, when bad weather threatens at lambing time, the gates of the loafing pen can be closed so that the sheep don't wander out and have their lambs too far from the barn.

OTHER SHELTER

If there is a scarcity of shade in the pasture, sheep should be provided with sheds or shelters where they can go in the heat of the day to get away from flies. The movable shelter shown here would have many extra uses during the winter as additional lambing quarters, ram pens, hospital pen, or storage shed.

USEFUL EQUIPMENT

LAMB CREEPS

Good-grade white pine and seasoned lumber removed from the barn are the best materials for constructing the equipment described below. Cheaper lumber will not stand up to the weight of the sheep. For the same reason, carriage bolts should be used in assembling gates and panels. Instead of using hinges on the gates, we have found it more useful to tie the panels with baler twine when two or more are needed, or to fasten brackets to each panel. Where panels are to be set up to remain in the same place for some time, they can be joined by boring a hole and threading a twelve-inch strand of heavy wire through each hole to make a loop. A stake is then threaded down through all the loops.

Single panels are easier to carry around, and two or more can be assembled in different ways to fill a variety of needs. A lamb creep should be constructed to withstand a great deal of plain old pushing and shoving by ewes who are determined to get in to all that appetizing grain. The creep

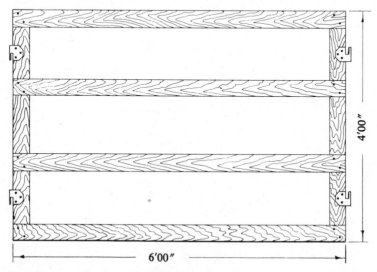

Pen Fence with Connector Brackets

33

Pen panel and bracket; each panel has the same piece (reversed) on both ends. Brackets can be made from 3″ x 3″ plates.

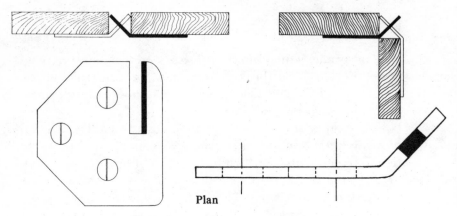

Elevation for Connector Bracket

Plan

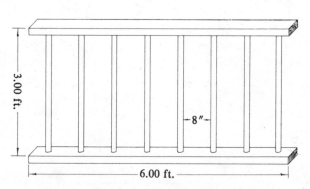

Panels for creep feeding.

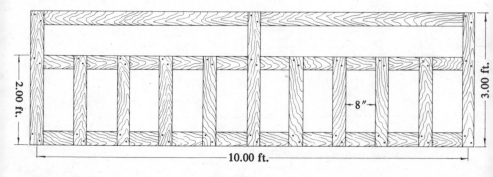

panel shown below is made with pine 1x4s and can be used as a creep gateway between two sections of the barn, as the front closing or a stall in the barn, or as the fourth panel of a separate enclosure. Allow three feet head space per lamb at the grain trough. The trough can be in the middle of the creep, with a space at one end for the lambs to walk around, so they can eat from both sides of the trough.

HAY RACKS

Here again, the material used must be strong. The five-sided hay feeder is useful both inside and outside the barn; it has the advantage of being easily moved and can be quickly turned upside down for cleaning. The rectangular feeders are good for permanant placement inside the barn. Individual hay and grain racks for ewes in lambing pens or the hospital pen can be bought from some farm equipment dealers or can be constructed from this plan.

GRAIN FEEDERS

A useful plan for a grain bunk for feeding ewes is given below. Rubber tubs, eight inches deep and about eighteen inches across, make good grain feeders and can substitute for water tubs if necessary. If not picked up just after feeding, these should be brushed out daily because lambs like to sit in them to get out of the wind on cold days. About five animals can eat from each tub. Water tanks, split down the middle, may also be used for feeding grain. The important thing is to have feeders that can be kept clean and that the sheep cannot stand in while others are trying to eat.

In the lamb creep, the grain feeders should be raised off the ground, or the lambs will play in them as if they were sandboxes, spoiling all the feed. Whatever kind of trough is used should be narrow enough to keep lambs from climbing in and shallow enough to be easily cleaned.

SALT AND MINERAL BOXES

Both indoor salt boxes shown below are two-compartmented, one side for salt and the other side for minerals.

Five-sided hay feeder.

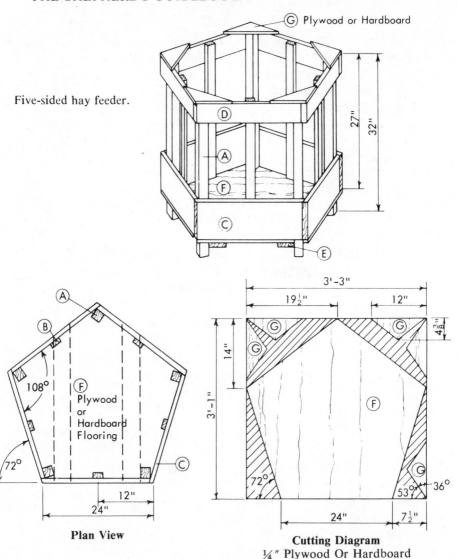

Plan View

Cutting Diagram
¼ " Plywood Or Hardboard

CUTTING LIST

Item	No.	Description
A	5	2x2 x 32 "
B	5	1x2 x 27 "
C	5	1x8 x 24 "
D	5	1x4 x 24 "
E	2	1x4 x 32-½ "
F	1	{ cut from 3'-1 " x 3'-3 " x ¼ "
G	5	} plywood or hardboard

Individual hay rack, for ewes.

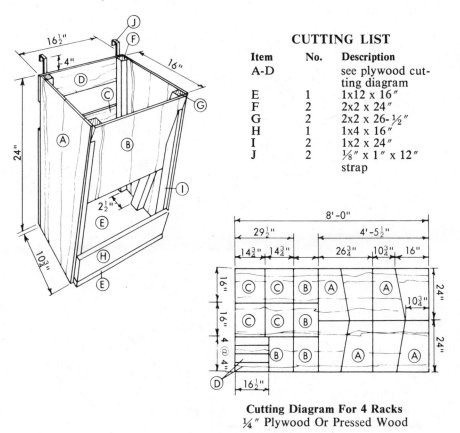

CUTTING LIST

Item	No.	Description
A-D		see plywood cutting diagram
E	1	1x12 x 16″
F	2	2x2 x 24″
G	2	2x2 x 26-½″
H	1	1x4 x 16″
I	2	1x2 x 24″
J	2	⅛″ x 1″ x 12″ strap

Cutting Diagram For 4 Racks
¼″ Plywood Or Pressed Wood

(Reprinted with permission from MWPS-3, *Sheep Housing and Equipment Handbook*, Midwest Plan Service, Ames, Iowa 50011.)

Sheep need salt at all times, preferably loose so that they do not wear down their teeth by chewing on blocks. The base of the indoor feeder extends a few inches to keep the sheep from standing close and thereby contaminating the contents with droppings. If the feeder is to be freestanding, the base should extend all round. The outdoor dispenser is mounted on skids to facilitate moving, and it may be suspended from a fence or building or stood in a dry area of the pasture. The roof should be waterproof, and the whole outside treated with creosote or commercial wood preservative.

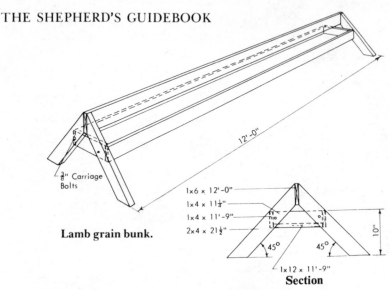

Lamb grain bunk.

1x6 x 12'-0"
1x4 x 11¼"
1x4 x 11'-9"
2x4 x 21½"

45°   45°

10"

1x12 x 11'-9"
Section

⅜" Carriage Bolts

12'-0"

(Reprinted with permission from MWPS-3, *Sheep Housing and Equipment Handbook*, Midwest Plan Service, Ames, Iowa 50011.)

Old rubber tires can be cut and turned inside out to make free salt or grain feeders.

LAMBING PENS AND GATE PANELS

You just can't have too many fence panels on hand; with them you can make chutes, pens, gates, sides for the pickup, and even temporary fencing to enclose a corner of the pasture. The value of such a temporary corral was well illustrated some years ago on this farm when five newly purchased sheep became panic-stricken during a bad thunderstorm. Instead of heading for the barn with the rest of the flock, they went charging off through the fence and across a neighbor's farm into the swamp. After about five days we found them and were able to get them up to a small patch of grassy meadow by the roadside, but no amount of cajoling, coercion, or cussing would persuade them to go up the road or through the woods to get back home. A roll of snow fence, conveniently stored nearby by the highway department, provided the answer—we neatly encircled the sheep with it so that they could be loaded on the truck for the ride home. Since then we have always kept a supply of fence panels ready for making emergency pens.

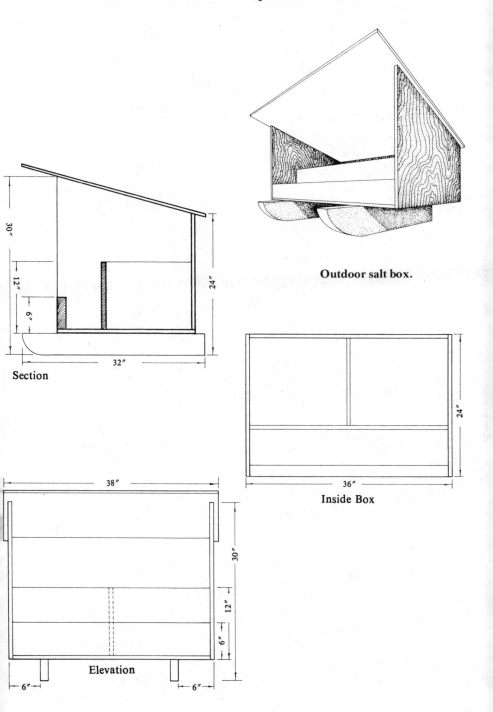

Outdoor salt box.

30″

12″

6″

24″

32″

Section

38″

30″

12″

6″

6″

6″

Elevation

24″

36″

Inside Box

39

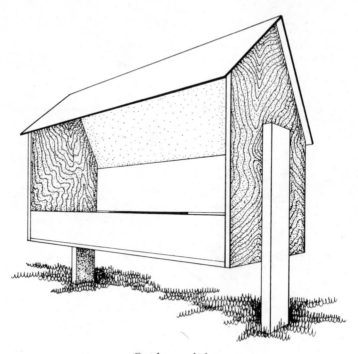

Outdoor salt box.

CUTTING GATE AND SORTING CHUTE

This can be a handy piece of equipment for separating lambs from ewes at weaning, for controlling sheep at spraying and worming time and for sorting flocks at breeding time. It can be constructed from plans in *Sheep Housing and Equipment Handbook*. When worming or sprinkling, a cutting gate at the end of a chute made from panels will suffice.

EQUIPMENT CABINET, RACKS

"Never put something down where you don't want to pick it up" is the word for storing equipment. Put all medicines in a safe place, and keep shears, scissors, knives, and other frequently used implements where they are readily available. It's very wearing to the temper to have to look for something like the iodine bottle at 2 A.M. Discarded milker inflations make useful covers for sheep shears.

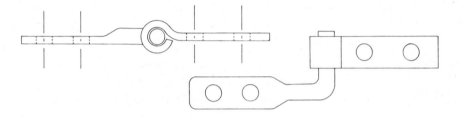

Smithing a hinge. A 6-inch piece of welding rod is heated on one end and bent to form a right angle. The other end is then heated, flattened with a sledge, and drilled. The socket is $1'' \times \frac{1}{4}'' \times 6''$ mild steel, heated, bent on an anvil, welded, and drilled.

Wall box for tools, medicine, etc.

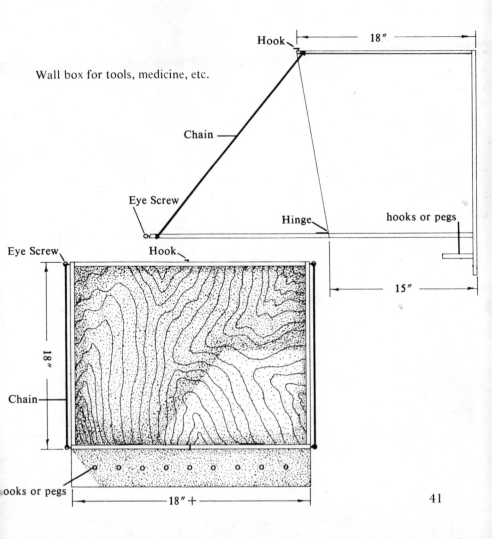

41

WOOL BOX

The wool box should be made of heavy plywood and strong hooks, and the pieces joined by two-inch strap hinges. For the shearer or his helper who is not too expert at tying fleeces, a wool box can help make a much neater bundle of wool.

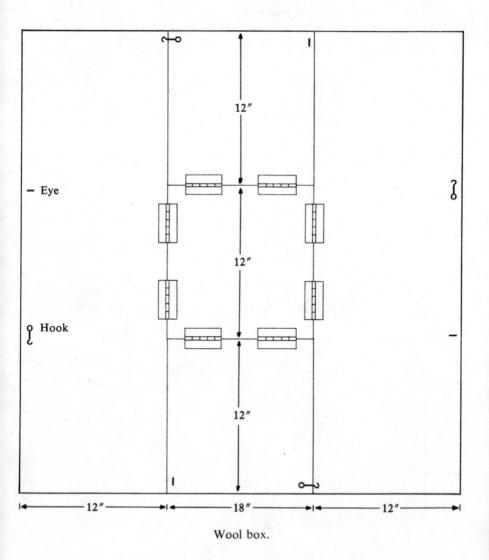

Wool box.

FENCING

A strong woven wire fence with a fairly small mesh is required to keep sheep in and dogs out. It should be at least forty-five inches high (even a rotund ram in full fleece will clear a fence from a standing position if he sees a low spot). Such fencing is expensive, but cheaper fence with a large mesh is a waste of money, as the sheep will learn how to work their way through it in no time. To keep dogs out, a single strand of barbwire along the ground on the outside of the posts, and two strands of barbwire, one six inches and one twelve inches above the fence, provide protection in most cases. This seems to work better than very high woven wire fences because a determined dog can climb such a fence if he wants to do so.

Even if it should be too expensive to fence in the entire sheep pasture in this way, it is strongly recommended that at least an overnight holding area be protected. The sheep can be allowed freely in the pasture during the day when the risk of attack by dogs is less than during the hours of darkness.

Whether to use wood or metal posts will depend on availability and local prices. Osage orange, black locust, chestnut, red cedar, black walnut, mulberry, and catalpa

A good fence design for sheep.

make the most durable wood posts, usually lasting fifteen to thirty years, and longer if treated with creosote. *Lee's Priceless Recipes* of more than eighty years ago suggested, "Take boiled linseed oil and pour it into pulverized charcoal to the consistency of paint. Put a coat of this over the timber, and there is not a man that will live to see it rotten."

Large posts last longer than small ones, and the diameter should be at least five inches with heavier posts at the corners. No matter what kind of posts are used, they should be well and firmly set into the ground.

ELECTRIC FENCE

Because of their wool, sheep are hard to control with electric fence unless they are introduced to it immediately after shearing. Hanging aluminum pie plates on the fence will arouse the sheeps' curiosity, and, having once had a jolt when they investigate, they will leave the fence alone. For dividing pastures for rotation, electric fence is usually quite satisfactory. Such a fence should have the bottom strand about twelve inches from the ground and the top strand twelve inches above this. Stakes with insulators are placed twenty to twenty-five feet apart, and the whole thing activated by a heavy-duty fence charger. One portable charger

Electric fence.

will usually last one season if it is switched off when not in use. A fence of this sort will keep out most dogs and has the added advantage that it is not permanent: the rather small posts that are needed are easily taken up and moved to another location.

SUPPLYING WATER

Sheep need clean water at all times. Ponds are not usually a good source of water, as the ground around the edge can too easily become contaminated with manure, and the trampling of their hooves at the waterline leaves a muddy area that attracts mosquitos. In many areas livestock are no longer allowed to use streams.

If the pastures are located near the barn a tank can be set up on a cradle and mounted on skids, and this can then be filled by a hose pipe. If a hose pipe is not practicable, the tank can be filled and then pulled into place on the skids. Galvanized cans of the garbage can variety can be filled and hauled out to the pasture with a garden tractor and wagon. An ordinary syphon tube will serve to transfer the water from the cans to the pasture tank. Such tanks should be on well-drained ground and constructed so that the sheep do not foul them with droppings and so that young lambs cannot fall in. Old-fashioned cast-iron kitchen sinks can be set up on blocks to make waterers, and cylindrical gas and water tanks, split lengthwise down the middle, make both feed and water troughs.

Where winter temperatures fall below freezing in winter, some form of heating will be needed. A small container fed by a pipe and float valve will work very well for an indoor waterer; for as many as 150 sheep we have used a $10'' \times 12'' \times 6''$ vegetable crisper from an old refrigerator with a small float valve attachment. The water stays fresh because the sheep are constantly drinking, and very little water stands in the container when it is not being used. In the winter the pipes are wrapped with heating tape, thermostatically controlled and secured to the wall to keep the sheep from pulling it loose, and the pan is set on packed-down dry manure, the warmth of which prevents the water from freezing in all but the coldest weather. If no running water is available, tubs of water can be set in the corners, with molasses added in cold weather to lower the freezing point. When larger containers are necessary, the water can be maintained above freezing by the use of an immersion heater, available from livestock supply companies.

HANDLING MANURE

The manure produced by healthy sheep is dry and pelleted, and is produced in seemingly vast quantities when they spend any length of time in the barn or loafing pen.

It is not necessary or desirable to clean the barns every day or even every week. At the beginning of the lambing season the barn should be thoroughly clean and bedded about six inches deep with clean straw. As the sheep eat hay from the racks they will discard the tougher stalks that they do not like, and these will collect on the floor and add to the bedding. Manure is warm, and the accumulation of manure and bedding form a warm floor for the barn that provides a comfortable place for the sheep in cold weather. In late spring, the whole place can be cleaned out by a tractor-mounted loader or scraper in the case of open-sided buildings and by manpower if the sheep are in a closed barn. This procedure is only a guide and not a rule. If the bedding becomes wet and foul, the barn must be cleaned more frequently, and filth must not be allowed to accumulate in the lambing areas.

In summer the sheep will go into the barn to get away from flies if they can. This often becomes a problem because the ingestion of large quantities of green grass has a predictable effect on the consistency of the manure, making it much harder to keep the barn clean. A layer of sawdust under a thin bedding of straw can be put down and scraped out as necessary.

Manure spreaders are expensive pieces of equipment, and for all but really large operations a small wagon will serve quite well to haul away the winter's accumulation of bedding. A nearby nursery garden or mushroom grower may have a ready use for manure. Don't store manure in piles for very long or flies will become a problem.

If the water pipes run along the floor of the barn, a layer of manure over them in winter, covered by dry straw, will provide enough warmth to keep the water from freezing.

Border collie.

SHEEPDOGS

There is no piece of "equipment" more valuable than a well-trained sheepdog. Such an animal can do the work of three or four humans. It takes patience, persistence, and a real love of dogs to train a sheepdog, but the result is well worth the effort. Border Collies, Australian Sheepdogs, and English Shepherds are all used in the United States, the most familiar being the Border Collie. Other sheepdogs, the Puli, Corgi, and German Shepherd, can also be trained to make use of their natural instincts.

Some breeds, Great Pyrenees, Komondor, and Kuvasz among them, are sometimes trained to guard the flock from predators. These dogs will live with the sheep and watch over them constantly.

47

3

Breeding

The shepherd's year begins with readying the flock for the start of the breeding season, which takes place any time from the end of July until the end of December, depending on locality and breed or type of sheep.

A flock of sheep, once comfortably established on a farm, will follow a predictable timetable of coming into heat and lambing. Even if other ewes are introduced, they will probably not come into heat until the main body of the flock does so. The onset of heat, or estrus, occurs in late summer, when the nights are cool, and even though the ram may have run with the flock all summer, the ewes will not breed until this time (except in the case of the Dorset, Merino, and Tunis breeds). Some of the earliest breeders are the Down breeds: Hampshires and Southdowns often have their lambs before Christmas, while the Hill breeds such as the Cheviots will not be ready to mate until October or later.

Once the ewes start coming in heat they will continue to do so approximately every seventeen days, for about three days at a time, until they are bred. Fertilization is most likely to take place if the ewe is bred towards the end of the heat period. The behavior of the rams will change at this time too. Instead of standing in the shade busily doing nothing, they will be out among the ewes, lifting their heads to catch the scent of a ewe in heat, and becoming more aggressive and bad-tempered.

PREPARING THE EWES FOR BREEDING

Gather the flock into a pen or yard, and give them some hay and water. When they are rested and quiet, look over every sheep for signs of sore feet, sore eyes, dirty tags of wool on the ewes' hindquarters, and an unthrifty appearance that would indicate worm infestation. Take the ram (or rams) out of the flock and put him in a comfortable stall in the barn.

Crutching is the light trimming of the ewes round the dock and udder, done at this time of year to clean off any dirty wool or tags that might interfere with the ram's serving them. This will need to be done again a few weeks before lambing.

It is important to worm to ensure that the ewes are in top condition, as ewes that are worm free at the time of flushing have a higher rate of conception.

Flushing involves putting the ewes onto higher quality pasture, or if none is available, feeding them about one pound of grain daily for a few weeks before breeding time. Oats, wheat bran, or oats mixed with corn are satisfactory, and the sheep will also relish pumpkins, broken open and scattered. The purpose of this extra feeding is to have the ewes on a rising plane of nutrition at breeding time, and is said to result both in the ewes breeding over a shorter period and in a greater incidence of twin births. Fat ewes will not benefit from flushing, as ewes that are overfat have trouble conceiving.

The value of flushing is said to be due to the increase of vitamin E from the improved feed or pasture. This vitamin is associated with fertility, and a lack of it may cause white muscle disease in lambs.

PREPARING THE RAMS

The rams, meanwhile, deserve some attention before the start of their busy season. Inspect the feet and see that the hooves are well trimmed (see foot rot under *Diseases*, chapter 6). If there is a lot of wool underneath the animal, this will need to be trimmed away in front of the pizzle and over the scrotum; high temperatures reduce fertility, and a lot of wool on the scrotum may result in the rams failing to settle some of the ewes. As a general rule, one ram will settle thirty ewes, while a yearling or an older ram will serve twenty or so. The ram can breed many more than this, but if he gets tired or down in condition the breeding season will be greatly extended.

Ram lambs can breed as young as five months, so any ram lambs from the previous year's flocks should be taken out before the beginning of the breeding season. Not only is it possible that they may breed their own mother and sisters, but they will also be competing with the flock ram for the ewes' favors, and the resulting fights will disturb the whole flock.

Some rams breed the ewes only at night, or when undisturbed, and it is quite possible to go through a whole season without ever seeing him work and still have every ewe deliver at the appointed time. To be sure that each ewe is serviced, the ram must be marked on his brisket (behind the breastbone and between the forelegs) with colored axle grease. Coloring pigment powder can be bought in a paint store and this, mixed with the grease, will leave a mark on the ewes' rumps when the ram mounts them; housepaint is undesirable, as it will spoil the fleece, while the powder and grease will wash off before shearing time. Now the fact that he mounts them does not necessarily mean that they are bred, and it would be very discouraging to find in January that none of the sheep were pregnant; so at the end of three weeks the color of the raddle, or mark, is changed. If all the ewes are seen to be coming back to the ram every three weeks, he is sterile and must be replaced. Sometimes only a few of the ewes fail to settle, and if another ram is available, they should be put to him before they are condemned as nonbreeders; even sheep should be allowed their preferences. The gestation period is from 144 to 147 days.

The ram should be in good condition, but not fat. A ram

that has just come from the show circuit has spent the past several months in pampered luxury, being fed choice food, sheltered from bad weather, and spared any hard exercise. He must be "let down" before being expected to go out and start serving the ewes. Some years ago we bought a beautiful ram at a livestock sale and brought him home to a comfortable barn and about twenty-five eager ewes. Hay was put out every day and the water fountain ran with sparkling fresh water—and that's exactly where the ram stayed. When the ewes went out he took up the best spot at the feed rack and munched all day, never venturing out into the pasture unless chased down there, and then coming back the minute the heat was off.

Desperate measures were in order. The barnyard gate was closed and the whole flock put in a pasture with a daily hay feeding. At first he ambled around after the ewes and even followed them into the woods when they went looking for late season grasses, but this was his undoing. He tried to climb over a fallen log and got stuck halfway, and there he stayed for two days, until we found him. The second time this happened we gave up and brought him back to the barnyard, penned him with two ewes for company and put another ram with the impatient flock. In about five months, one beautiful chubby lamb was born. The other ewes had their lambs several weeks later—and one handsome ram went off to be someone's pet.

The moral of this story is not that a ram for a show is not a good buy, but rather that if such a ram is highly fitted he must be prepared for the hard realities of life before he can be expected to go to work.

BRINGING THE RAM TO THE EWES

When the ewes have been on the flushing for about two weeks, the ram can be put in with them. During lambing time you will need to spend considerable time with the sheep, and it is therefore to your advantage if all the ewes can be bred to produce their lambs within a close period.

The number of ewes any one ram will serve depends in part on the system used in putting the rams with the ewes. The ram, or rams, may run continually with the ewes, being separated into flocks at breeding time; or, the rams may be taken out by day and rested in the barn, which will probably enable each ram to serve more ewes than if he were using up energy during the daytime; and a third method is to use a teaser, or vasectomized ram, to indicate which ewes are in heat, and to bring these to the ram when they are ready. In some cases the teaser is a normal ram with an apron tied around him to prevent service. This will work for a little while but then the ram is likely to damage his sensitive male organ and to tire, and he then will miss some of the ewes that are in heat.

The first method is the most natural one. Nature plays a part in a successful breeding program, and the ram and ewes in each flock will be quieter and more contented if they are not continually disturbed. Taking the ram away by day may rest him, but many rams become upset and irritable when separated from their ewes, and the ewes, too, like to stay close to the ram at the time they are ready to breed, becoming restless when separated from him.

There is, of course, another method. This is the winner-takes-all system where all the rams and ewes are left together, and may the best man win. In theory, each ram breeds some of the ewes as they are ready, but in practice the strongest and best ram is usually so busy fighting off the competition that he is too tired to breed some of the ewes while at the same time preventing the other rams from doing so. The ewes are disturbed by all the pushing and fighting, and when the lamb crop does come, no one knows who begat who. To avoid breeding them to their brothers and fathers, the ewe lambs of the crop must be put to a new ram the next year.

In the breeding of purebred sheep, great improvements may be made by close breeding, the closest possible mating being that of full brother and sister. This could concentrate the most desirable traits, but it is just as likely to double-up bad qualities, resulting in decreased size, strength, and fertility. When the flock is of mixed bloodlines, the close mating of family strains can have equally undesirable results. It is recommended that inbreeding be left to those purebreeders who have made an extensive study of genetics and can predict results of their breeding program.

If the flock is large enough to warrant more than one ram, these may all be run together with the ewes during the summer. They will probably get along quite peaceably until their built-in calendar tells them that mating time is here. When this happens they must be separated, or they will start fighting and may injure each other. Not only will they fight with each other but they may become more aggressive towards humans—it is not wise to turn one's back on such an animal. A bell round the neck will signal when a ram is coming and will also help in finding him and his ewes on foggy mornings.

The fences and gates dividing each ram and his ewes from the other rams and their ewes should be very strong. Rams can clear a four-foot fence from a standing start. Our Cheviot ram was undaunted by a four-foot high, two-foot thick barnyard wall, and it took two people to rescue him from the top. The same season, a Hampshire ram, failing to make it over the top of a gate made of 1 × 4s, tackled the obstacle head-on and ended up in the sheep hospital with a very sore head.

If the different flocks can be some distance from each other, so much the better. The rams will be less inclined to walk up and down the fence line, guarding their territory, and there is less likelihood of their breaking through.

REARING GOOD MOTHERS

Bred ewes need shelter, exercise, water, and feed. While the weather remains warm they will do well on pasture and will benefit from the daily exercise. When the breeding period is over, ewes and ram can be put back on normal daily keep, with hay supplemented as the pasture becomes thin and dry after frost. Even after the pasture is dormant the ewes will enjoy wandering around nibbling here and there at odd greenery they find. This is good for them, as long as they pick up nothing poisonous, like wilted wild cherry.

Ewes that are closely confined in a barn or yard all the time are liable to suffer from constipation, one of the worst enemies of the pregnant ewe, and tend to produce sickly lambs.

Ewes can stand almost any cold as long as they are dry and well fed, and will often choose to stay outside during a snowfall. Sleet and driving rain can penetrate the fleece, and the animals should have access to shelter during this kind of weather.

PREPARING FOR LAMBING

About two months before the lambs are due, the wool should be shorn from around the dock (tail) and udder and

Champion Columbia ewe.

from the inside of the hind legs. Great care must be taken not to injure the teats, especially when trimming maiden ewes whose teats are not yet developed and therefore easily nicked if hidden in the wool. This trimming, or crutching, will make it possible for you to watch the development of the udder, an indication of how far pregnancy has advanced—it will become pink and congested when lambing is imminent. At the same time, the udders of the older ewes can be checked for signs of injury or old scarring, which may make it impossible for them to nurse their lambs properly. Occasionally a ewe has only one side of her udder in working condition; she will be able to feed a single lamb quite well, but will need some assistance in raising twins.

Cleaning around the tail makes the process of lambing more sanitary and makes the shepherd's job easier if assistance is needed. When the vulva is free from manure and fleece, the shepherd can look in on the ewes and check on the progress of birth without disturbing the animals. This is important because the ewe likes her peace and quiet at this time, and will sometimes fail to take proper care of her lamb if she is constantly disturbed.

No ewe should go through pregnancy bothered by pests, either internal or external. At the end of the summer a fecal sample from one or two of the flock should be taken to the veterinarian for a worm check, and if worms are found, appropriate medication must be given (see chapter 6). If the wool is worn off in places and the animals are constantly rubbing themselves, they should be inspected for evidence of ticks. Sheep ticks are about the size of a fly and can easily be seen on the skin. The hard, dark, brownish red pupa cases will be found in the wool. A few ticks at this time can become a horde by spring, and if left untreated the pests will drop off the sheep and go onto the young lambs, possibly causing severe anaemia. This is no time for rough handling: the dusting or spraying should be conducted as quietly as possible, using any veterinary-approved spray or powder.

Planning for "the shepherd's harvest" would not be complete without a few words on the shepherd's own preparations. Assemble in a carrier or clean box some clean soft cloths and towels, a pair of scissors, jar of Vaseline, piece of soap, flashlight, and baby bottle. The hole in the nipple of the bottle must be enlarged a little with a hot needle. Another box to serve as a "warm box" for a chilled lamb may also be useful. It should contain a hot water bottle and plenty of soft cloths or old pieces of sheepskin. An infrared heat lamp or two will be a lifesaver for lambs born during very bad weather; be sure to get the red ones and not the white floodlight variety. A supply of pails for water in the lambing pens will be needed, as well as some grain pans (clean discarded dishpans work very nicely). To help chilled or weak lambs, have on hand a supply of honey, a 10 cc. syringe, and a bottle of 50 percent glucose, obtainable from a veterinarian.

If the barnyard or loafing pen is large, the flashlight should have a really powerful beam that will reach every corner: you are less likely to miss a ewe who has decided to have her lamb in the most inaccessible place if you can see everything that's going on; and the new lamb that has wandered away while its mother cleans up her twin has more chance of being found.

This may seem like a lot of equipment, but when you need it you need it in a hurry. The sheep will not necessarily have their lambs in tidy intervals of two or three a day. One day you may have six new lambs and the next none, or things may be routine and dull for a week and then one day a ewe will have triplets, a young ewe may refuse to nurse her lamb, and another ewe may decide to produce her young outside in a blizzard. So, be prepared.

4

Lambing
and
Flock
Management

As lambing draws nearer, you should be spending more and more time with the flock, just watching them. Look for any variations in their normal behavior, symptoms that might suggest one is ill, or other occurences that cannot be described in words but will be apparent to an experienced eye. There is no time when your attention is of more value and your presence more important, and you must be ready to give assistance where needed and to see that the new lamb gets the best possible start in life.

During the last five or six weeks of pregnancy the fetus will have made about 70 percent of its growth. The ewes will have become rounder and heavier and more inclined to relax in the barnyard than to go out in the fields. Most expectant mothers feel the same way, but nonetheless, daily exercise is a must if the lambs are to be strong and healthy. Such exercise should be gentle, with no chasing and no jumping over obstacles. When the sheep come back to the barn to rest and chew the cud they should have a place where they can settle on dry ground. Sheep don't like mud and will stand for hours rather than lie down in it, and when they do eventually lie down, not only is there a danger of their catching cold but they become dirty and present a very unattractive lunch counter to the new lambs.

About a week before her lamb is due, sometimes more, the ewe develops a discernible udder that turns bright pink immediately before parturition. There may be a slight discharge from the vulva, which will appear red and swollen, and the body will appear to be sunken in front of the hips. When she is ready to give birth, the ewe may wander off by herself, ignoring her feed and pawing at the ground as if preparing a nest for the lamb. Ewes about to lamb look restless and ill at ease, and sometimes they baa as if calling to the lamb—or maybe calling for a little sympathy from their friends. They should be allowed, as far as is practical, to pick their own place, and should be left to walk up and down, or to settle there, undisturbed. If the spot chosen is obviously unsuitable, such as a place in the middle of a yard occupied by other livestock, or near a deep puddle or other hazard, she should be moved to safety, urging her along gently and only just as far as is necessary for safety; once ewes have chosen a place, they don't like to change. A ewe may deliver her lamb or lambs quite successfully without displaying any of these signs; most likely you will come into the barn one morning ready to feed the sheep and find a ewe standing proudly beside her new offspring, having accomplished the whole thing, as sheep have done for thousands of years, without any help from anyone.

THE BIRTH

When the water bag appears the ewe will start laboring in earnest, lying down to strain then getting up and changing her position. This may go on for half an hour before the lamb's nose appears, followed by the front feet as the ewe continues to strain. Within a short time the whole lamb will slide out. The ewe will turn around and start enthusiastically licking her offspring, crooning to it and nudging it. This crooning mothering sound is distinctive and, if you are making the rounds of the barn or yard on a dark night, you will be able to tell if a ewe has a new lamb by listening for her talking to it. In the same way, you will be able to recognize the characteristic grunting and straining sound of a ewe in labor.

In a short time the lamb will struggle to its feet and start looking for a meal. The ewe will lick its tail and encourage it to move in the right direction, while hunching slightly to make the nipples more readily available. If a twin is on the way it may come immediately, or it may not appear for half an hour or more. When the ewe has her first lamb dried off, put mother and child in a lambing pen with a pail of tepid water and some good hay, check the teats to see if they are blocked (if so, remove the wax plug by rolling the teat between thumb and forefinger). Then leave her undisturbed. As a precaution against infection, the navel can be dabbed with a swab soaked in iodine when the sheep is relaxing over her hay. The warm water encourages the ewe to drink copiously and by doing so to expel the afterbirth more easily.

Leave the ewe and lambs in the lambing pen for three or four days. If so many lambs are born at one time that the lamb pens are all full, take from the pens the mothers which show the strongest mothering abilities. But before letting them go, mark mother and young with some sign of identification (using sheep paint or crayon) so that if they become separated you will know who is missing. This also helps in matching up mothers and lambs if one or other should become sick or injured.

If the ewe has lambed outdoors and the weather is very bad, the mother and lambs will have to be brought under cover as soon as the lambs are born to avoid chilling. To get the mother to follow you, pick up the lambs, using an old towel because they will be wet and slippery, and walk slowly backward into the barn so that the ewe can see the lambs all the time. Keep going until you are in the lambing pen. Carry the lambs by the whole body, never by the front legs. Some nervous ewes keep running back to where they dropped the lamb, and it's vital not to get excited and hurry the process. Otherwise, the ewe may panic and run back into the flock and you will have all sorts of trouble locating her and reuniting mother and young.

Occasionally a ewe with an extra strong mother instinct that is about to lamb will steal a lamb from a ewe that has just given birth. When you find two ewes standing by one or two lambs, you can sort this out by checking the ewes' hindquarters; the ewe which has just lambed will have traces of bloodstained membrane clinging to her.

If the flock has been well nourished and the ewes have developed well-filled udders, very little attention will be needed.

PROBLEM CASES

The foregoing instruction covers a normal lambing, but in every flock there are always some problems. First, it's important to understand the normal processes of birth and to recognize the need for calm and quiet, and of course, cleanliness.

A routine presentation consists of the appearance of the nose and feet, with the pads of the feet pointing down, and then the whole head. Sometimes only one foot appears, and this may cause trouble. If the ewe continues to strain but makes no progress it may be that the other foot is turned back and wedged against the pelvic girdle. Head the ewe into a stall, or place lambing hurdles around her to make a pen. Have everything at hand that you might need—soap, warm water, Vaseline, uterine pessary to counteract infection, clean towels, and an assistant if one is available. There is some disagreement among veterinarians and shepherds on whether the ewe will deliver her lamb normally when only the head and one leg are presented, or whether she will need assistance. It is better to try to find out what the problem is than to let the ewe continue straining with no progress being made. First ascertain how many lambs are involved, since it is quite possible that there are two lambs wedged in the birth canal. Clear a place on the ground and put down a clean cloth. Right here comes the first problem: if the ewe is standing up, you have to get her lying down as gently as possible. When two people are doing this then the ewe can be tipped off balance and gently let down onto the cloth, but if only one is working, the ewe must be tipped off balance by holding the nose in the left hand, while standing behind her, and turning the head back over the shoulder until she goes down. The easiest way is to wait for her to lie down and strain and then to carefully but firmly put one hand across her and hold her down. So far so good—and now to wash the hands and arms. This shepherdess, who weighs less than many of the ewes on the farm, has eased this procedure by having nearby several towels wet with Lysol solution in a plastic bag; then when the sheep is lying quiet, the towels are handy for wiping hands and arms. Next, one of the towels can be used to wipe off the ewe's dock; if the crutching job was done properly, this will not be very soiled.

Rub a little Vaseline on the hand and make the fingers into a cone. Work the fingers gently into the birth canal and try to find the other leg. If there are two lambs side by side, push back the one that is the furthest in. Now find out whether it is a right leg or a left leg that is turned back. Have the ewe lying on the same side as the leg that is turned back so that the backward leg is uppermost. Pull on the extended leg a little, easing the head out by working the skin of the ewe back over the crown of the lamb's head. If you are doing this with the help of an assistant, one will be steadying the sheep while the other works on the lamb. But if alone, one soon comes to realize that nature should have given a shepherd four hands. Fortunately, most ewes will lie quietly if spoken to soothingly and occasionally steadied with an arm or hand. With the head and one leg out, twist the lamb a little by drawing the leg upward and the head downward, and the other shoulder will slip out. Pick up the lamb and put it under the mother's nose, wipe off your hands with the disinfected cloths, and insert a hand again to see if there is a twin waiting to be born. If you feel a nose, wait a few minutes and in nearly all cases the lamb will slide out easily; if it comes backwards, wipe the mucus out of its mouth (see the section on *backwards lamb* later in this chapter) and put the second lamb at the mother's head, together with the first.

This same method is used when a lamb is too big for the ewe to deliver without assistance. It is tempting to try to draw out both feet at once, but this means that the head is coming through the pelvic girdle at the same time as the shoulders, which are the widest part of the lamb's body. By bringing but one leg at a time, the length of the body is extended and is thus narrower.

In this case and in all others of assisted birth, do what you can, but if the problem is too difficult, call for help before the ewe is exhausted by fruitless laboring. If the veterinarian cannot be located, or cannot come for some time, keep the ewe quiet and stay with her. Remember that it is better to be on hand if she eventually delivers a half-dead lamb and is herself totally exhausted than to get overanxious and permanently injure the lamb and the birth canal by pulling too hard. If the veterinarian is unable to come but will see the ewe at his office, very carefully load her into the back of the station wagon or pickup and drive there as soon as possible.

Sometimes a lamb born after prolonged labor appears to be dead, but don't give up until you have tried wiping its mouth and nose and rubbing it briskly but lightly with straw or a rough towel to simulate the mother's licking. A light slap on the ribs or a quick dunking in cold water will help start it breathing. As soon as you have cared for the lamb, place it where the mother can see and smell it and give the mother ½ cup of coffee mixed with 2 aspirin, 2 tablespoonfuls of honey, and 2 tablespoons of whisky, dribbling it into her mouth through a baby ear-bulb syringe. If this is your first time at such obstetric complications you may find this pickup a great help to the shepherd too.

Don't be in too much of a hurry to get the sheep on her feet, but let her rest for a while. Pretty soon she will sniff her lambs, and when she starts licking them this mothering and cleaning up will do wonders to revive her. If it is very cold the lambs may be given an ounce or two of warm milk from a baby bottle to sustain them until the mother feels like taking on the job. It is a mistake to take the lambs away to be cared for somewhere else until the ewe is up, as they need each other and the mothering bond should be broken only in cases of great emergency, such as the onset of freezing rain.

To prevent the lambs from losing their vital body heat, put them on a thick bed of straw and cover them with cloths; a well-wrapped hot water bottle will help keep them warm. As soon as the mother can be moved, get them all into a clean, comfortable pen.

When the head is showing but no further progress is being made, push it back again through the bones. Slide the fingers along the neck and shoulder, hook the leg with one finger and draw it forward, straighten the leg at the knee, and then draw the foot out. From there, proceed as in the previous case.

If the ewe cannot deliver the lamb because the feet are out but the lamb's head is turned, have the ewe lying so that the lamb's head is down and turn one leg back. The lamb should now come out easily. If not, push it back a little and turn the head. When drawing the lamb out, pull slightly downward, towards the udder, and work with the mother, pulling when she pushes.

Sometimes, however, it is very difficult to get the head out if the forefeet are showing and the head is turned back. In such cases, get two pieces of clean soft string, make a noose on the end of each piece, and slip a noose over each leg at the ankle. Then push the legs back into the womb, making sure that the string protrudes. With the lamb's shoulders no longer filling the neck of the womb, it should be possible to use one hand to draw out the head while the other hand brings along the front legs by pulling on the string. Once the lamb is safely out and it is ascertained that there is no twin waiting, a uterine pessary should be inserted into the neck of the womb to combat possible infection.

BACKWARDS LAMB

When the lamb is coming backwards (the pads of the feet are facing upwards or the tail shows) it is often found that the lamb sticks at the ribs. If the feet show, hold the legs and gently work the lamb out with a seesawing motion. If only the tail shows, push the lamb back a little and hook the finger over the hock joint, drawing out first one leg and then the other. When a lamb is coming backwards its head is lying in the fluids in the birth canal, and it is therefore necessary to work as quickly as possible so that the lamb will not drown. As soon as the lamb is out, wipe the mucus from its mouth and pick it up by placing it between your forearms, supporting the head with the fingers of one hand and the shoulders with the other hand. Then swing it slowly back and forth a few times to allow the mucus and fluid to drain from the throat and lungs. Don't pick up the lamb by its hind feet. Put the lamb by its mother, as described before, but don't go away. There will very likely be a twin which may also be coming backwards, and the mother is going to be too exhausted to get up and see to it. If the lamb is presented breech first, push the lamb gently forward, then hook a finger behind the hock and gently draw back the leg, keeping the leg bent, and then grasp the foot and draw this back. Deal with the other leg in the same way. If both the lamb and your hand are small, it is sometimes possible to draw out the lamb without straightening the legs, but this should be done as a last resort.

Occasionally a lamb will suffocate in the fetal membranes if these fail to rupture at birth. When a lamb is seen to have what looks like a plastic bag over its head you should immediately break it to permit the lamb to breathe. If a ewe is seen to be straining to deliver a lamb, and the visible part of the head is covered with the membrane, lambing pen hurdles should be set up to prevent the ewe from dashing off, and the film over the nose should be broken. Leave the hurdles in place, in case the lamb is weak and the mother and young need more attention.

In summing up, assistance at lambing time should be based on several guidelines. Cleanliness and careful handling are of prime importance. The cause of the trouble should be ascertained before any attempt is made to correct the situation. Be sure that the ewe really needs help: labor sometimes takes several hours, and if the sheep is walking around, making a nest in the bedding, answering the calls of other lambs who may be near the barn, or actively straining, then be patient—as much harm can be done by rushing in too soon as by waiting too long.

There are some situations that call for expert assistance. Don't pull on any part of a lamb until you are sure where and how the lambs are lying in the birth canal. Never be rough,

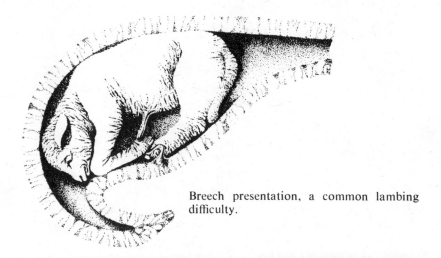

Breech presentation, a common lambing difficulty.

for if the uterine lining is torn, infection may set in and the ewe will be lost. Don't wait too long to send for assistance, but give the veterinarian a chance to save both mother and lambs.

Look out for the ewe who is lying in a dark corner or away from the other sheep: she may have lambed or may be in process of doing so. Watch too for an occasional ewe who may be walking around straining as if trying to have a bowel movement: she may be trying to give birth to a lamb.

THE FIRST TWENTY-FOUR HOURS

This is the most critical time in the life of the new lamb. As soon as the mother has started to clean it, the lamb should be able to get to its feet and find its first meal. The ewe may have a small wax plug in the teats, which can be removed by rolling the teat between thumb and forefinger. A smooth stroking motion between the fingers, downwards, will usually bring the milk down. This milk is rich in the antibodies needed to protect the young lamb against infection, and the lamb should get its first drink within an hour of birth. If the milk has not come in, possibly because the ewe is exhausted or frightened, the lamb can be sustained with a few ounces of warm cow's milk fed from a baby bottle at hourly intervals, until it starts to nurse from its mother.

Sometimes a very young ewe will run away from her lamb after it is born and refuse to have anything to do with it if they are penned together, even to the extent of kicking it away. When this happens, put the lamb in a safe place with a few ounces of milk to keep it warm, and use a heat lamp if the weather is very cold. Put the mother in a pen with some hay and a bucket of warm water, and let her rest. In nearly all cases it is possible to bring the lamb to her after a few hours and have her welcome it quite happily.

The cord breaks about two inches from the navel at birth; if it does not, the mother will bite it. If a long piece of the cord should remain, however, this can be cut and the remaining piece disinfected with iodine. Another precaution at this time will save those twins which are born to old ewes, to ewes with poor milk supply, or to just plain ornery ewes. There is a tendency for the smaller, weaker lamb to wander away, or for the firstborn lamb to stray while the mother is busy cleaning up the twin. If the lamb is weak and the mother is busy with the twin, there is nothing to be gained by trying to force the lamb to nurse. Instead, put both lambs and mother in a lambing pen and then get the weaker one dried off and warm. Put about three ounces of cow's milk in a baby bottle and give the lamb a warm drink. Make the hole a little larger by pushing a hot darning needle through it. Most likely the mother will take proper care of both when she has had a good meal of some nice hay and a drink of warm water. If she does not, then other measures must be taken, about which more later.

FEEDING NURSING EWES

Well-nourished lambs grow quickly. It's good practice to feed the lambs through the ewes, stimulating the milk flow of the ewe by feeding rations high in protein and minerals. A good mixture is 6 pounds oats, 2 pounds wheat bran, and 1 pound of cottonseed meal or linseed meal. Give one pound a day, and in addition feed one to three pounds of good legume hay. A similar amount of silage helps to stimulate the milk flow, but this must be very clean and free from mold, or both the ewe and the lamb will suffer digestive troubles.

THE IMPORTANCE OF COLOSTRUM

The first milk of the ewe is called colostrum milk and is vital to the well-being of the lamb. This early milk may be as much as one hundred times richer in vitamin A than later milk. It is loaded with protein in a form easily assimilated by the lambs, and is rich in antibodies to protect the lambs against bacterial infection.

Lambs may miss their colostrum. Sometimes a young and inexperienced ewe, or one who has had a difficult labor, may run away from her lamb and refuse to let it have that crucial first meal. Or a larger twin may take all the milk. If the ewe has a painful udder or damaged nipples, she may reject the lamb. These lambs, deprived of their first meal, will be seen following the mother, trying to nurse and bleating with hunger. If not cared for and fed, they will go off into a corner, the body temperature will fall, and they will become weak and lethargic. If you see a lamb whose mother is not caring for it, put your finger in the lamb's mouth: a healthy, well-fed lamb will have a warm, almost hot, mouth; if the mouth feels cool, the lamb should immediately be given a few ounces of warm milk. Some colostrum milk may be drawn from another ewe that has just lambed, if the condition of the umbilical cord indicates that the hungry lamb is just newly born. If there is a dairy farm nearby a supply of cow's colostrum milk could be obtained and kept in the freezer. Most farmers will be happy to let you have a gallon or so, as the first milk is usually thrown away anyway. A supply can be stored in baby food jars, or even small plastic bags. The milk must have been taken from the cow within twenty-four hours of the birth of the calf in order for the colostrum properties to be any good to the lamb. Ewes' colostrum milk may also be frozen for later use. Milk has to be heated very slowly when thawing, because it is very thick and will clot if heated too rapidly. Artificial colostrum, to be used when none is available from a natural source, can be made from

24 ounces cow's milk, 1 beaten egg, 1 small teaspoon cod liver oil, and 1 large spoonful sugar.

This mixture can be fed six times a day, four ounces per feeding, for the first forty-eight hours, and then four times the third day and thereafter gradually mixed with ordinary milk so that the lamb doesn't have a sudden change of diet. As soon as the lamb is warm and eating well it can be left with its mother, who may either come into full milk later or may always be a poor milker (in which case she should be culled at the end of the season). The lamb will have to receive supplemental feeding until it is well started on hay and grain. Whether the mother is feeding the lamb or not, she will usually mother it, and the lamb is best left with her for warmth and comfort. It will grow better and will start nibbling hay and grain earlier if it has a mother to set a good example.

Some ewes will refuse to let their lambs nurse, and yet on inspection, nothing is found to be wrong with the udder or teats. In this case the trouble may lie with the lambs. Those first lamb teeth may be very sharp, and a short smoothing with a common nail file or emery board will correct the problem.

DOCKING AND CASTRATION

Both docking and castration should be carried out on a cool day when the sun is shining, so that the lambs will want to go outside in the fresh air. It is best to do the job early in the morning so that the lambs have all day outside, and so that they can be watched for bleeding. If the weather turns wet or very cold and snowy, defer the job, because lambs running outside immediately after the shock of docking will be susceptible to cold. If the job must be done indoors, the bedding should be clean and dry. Care must be taken not to overexcite the lambs.

All equipment used must be clean and sterilized. Surgical steel tools and hypodermic needles may be boiled in a pan of water for twenty minutes and then wrapped in clean wax paper until used. Clean blood and wool off knives and emasculatomes immediately after use, and take care to see that there is no rust. Always use fresh, sharp needles.

DOCKING

When the lamb is about a week old, and before it is two weeks old, the tail should be cut off, or docked. This tail, which looks so cute wiggling as the lamb nurses, becomes a reservoir of manure and germs during the hot summer months. If ewe lambs are to be kept for breeding, the tail will be a disadvantage during the mating season and a positive nuisance at lambing time. So get it off! Dock the ram lambs too, as they may have a period of scouring (diarrhea) when they are being fattened for sale or butchering, and then this messy tail must be painstakingly cleaned off to avoid the risk of fly strike and maggots.

A surgical steel tool called an emasculatome is used for docking and, if properly handled, will leave a clean cut with a fold of skin over it that will permit clean, quick healing.

The cut should be made about 1½ to two inches from the body, and the stub treated with Bloodstopper or any other antiseptic fluid. An injection of 0.5 cc. of tetanus antitoxin is a valuable precaution against infection from the tetanus spores which lie dormant on many farms, especially where there have been horses. For ewes injected with Fort Dodge vaccine two weeks before lambing, the tetanus antitoxin will not be necessary. If a knife is used for docking, a string should be tied around the tail above the cut to discourage bleeding; the lamb should be kept under observation for half an hour or so, at which time the string can be removed and the bleeding will have stopped. The stub should be dusted with Bloodstopper.

The bloodless, rubber ring method of docking has become widely accepted. An elastrator is used to place a rubber ring around the base of the tail and, if the lambs are to be castrated, around the scrotum. This cuts off the circulation, causing the tail to drop off in a week or so. Immunization against tetanus is necessary with this method also. The lambs so treated must be watched to see that there is not a rotting stump of tail to draw flies.

Docking.

Applying Bloodstopper.

Injecting tetanus antitoxin.

Lambs can also be docked with a hot chisel. The lamb is placed on a board on its rump with the tail extended along the board. Another board, with a notch cut to fit over the tail, is placed between the lamb and the chisel. The chisel should be heated to a full red; if it is too hot a severe burn may result, and if it is not hot enough, the blood vessels will not be sealed.

Adult sheep with long tails and/or lambs that have reached more than four weeks of age cannot be docked so easily. This becomes a surgical procedure and the assistance of a veterinarian is called for.

CASTRATION

To castrate a lamb, have a helper hold the lamb with its back against his body, grasping the legs firmly to stretch the rear legs upwards. Disinfect the scrotum and the knife with Lysol and cut off the end of the scrotum to expose the testicles. Grasp the base of the scrotum between thumb and forefinger, pressing against the lamb's body, and pull out the testicles and cords with the other hand. Cut them off, disinfect the wound, and dust with Bloodstopper.

Certain markets will deduct from the buying price if male lambs are not castrated. In fact, however, ram lambs grow out better and with less undesirable fat than wether (castrated) lambs; so, if the buyer does impose a deduction it is because he is using this as an excuse to pay a smaller price. There is no loss of flavor in young rams, but they should be separated from the ewes before they reach puberty, or the age of ambition, for otherwise they will chase the girls all day instead of standing around and growing. If ram lambs are to be kept more than one season for wool production, castration is recommended.

There is also a method of castration using a Burdizzo, a tool that crushes the cords. As no wound is made, the lamb becomes a cryptorchid, and thus combines the qualities of both rams and wethers. When this method is used, the lambs must be marked to indicate they've been treated, because it will likely be difficult to tell whether the animal is a ram or a wether.

FEEDING LAMBS

Lambs will begin to nibble on grain and hay when they are about two weeks old. By the time they are four weeks old they will eat four ounces per head daily, and when they are eight weeks old they will eat twelve ounces. Suckling lambs are more apt to respond to additional energy (carbohydrates) than to protein, because the dam's milk provides protein of high quality and plenty of it. Suggested creep rations are:

1.	Corn	50 percent
	Oats	30 percent
	Soybean meal	15 percent
	Molasses	4 percent
	Limestone	1 percent
	Antibiotic	15–20 mg per lb
	Vitamin A	1,000 I.U. per lb
	Vitamin B	200 I.U. per lb
	Vitamin E	20 I.U. per lb

(Feed with limited amounts of high-quality alfalfa hay. If the ration is pelleted, reduce both the corn and oats 10 percent each, and the soybean meal 5 percent, and add 25 percent alfalfa.)

or,

2.	Shelled corn	55 lb
	Oats	25 lb
	Oil meal	15 lb
	Alfalfa pellets	5 lb
	Hay	Free-choice

Antibiotic may be added at the level of 10 mg per pound of feed in started rations.)

Antibiotics added to the feed will usually increase gains and feed efficiency. They will lower disease level, especially the incidence of scours, enterotoxemia, pneumonia, and soremouth, and will result in a more uniform lamb crop. They are, however, not essential, nor are they a substitute for hygiene and good nutrition.

Lambs enjoy Calf Manna, or its equivalent in calf starter, sprinkled over their feed.

LAMB CREEP FEEDING

A lamb creep is an area set up where lambs may be fed and where the ewes cannot reach. It provides a method of giving supplemental feed to the lambs during the nursing period, helping to get them well started in life and increasing their resistance to parasites, infection, and colds. The lamb creep should be located in an area where the lambs will use it readily; choose the brightest and sunniest corner of the barn, or, if outside, a convenient dry area near the place where the ewes commonly come to rest and ruminate. A heat lamp over the creep will encourage the lambs to spend more time there. The ration in the creep need not be complex but it should be tasty and nutritious. Strict cleanliness of feeders and waterers is a must; at the end of the day the unused grain can be taken out and fed to the ewes later, and the waterers must be regularly cleaned and refilled daily. When the ewes are fed daily the lamb creep can be replenished at this time, as the lambs will follow their dams into the barn or feeding area and will be more contented in the creep when the ewes are nearby. Until the lambs are six weeks old the grain used in the creep should be cracked, crimped, or rolled.

Instructions for making lamb creeps are given in the plans in chapter 2. If only a few lambs are to be fed, a single stall of the barn can be readily converted by putting a barrier of up-right slats or rollers across the fourth side. These uprights should be eight to ten inches apart. The lambs will soon learn to go in, and if they are slow to start eating, a little brown sugar sprinkled over the grain will tempt them. They are great imitators, so once you get one of them going the others will copy. If lambs are to be fed a high-energy ration for early and quick gains, they should be vaccinated for entero-toxemia, but if the creep feeding is a supplement to pasture or hay feeding, this problem is unlikely to occur.

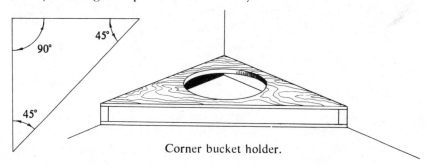

Corner bucket holder.

SOME COMMON LAMBING PROBLEMS

ENTEROTOXEMIA (OVEREATING DISEASE)

Overeating disease, known also as pulpy kidney, affects sheep of all ages, even very young lambs if the mother is a heavy milker, and can be a problem among newly weaned lambs in pastures or feedlots. Usually it is the large, single, and most vigorous lambs that are lost. Husky, well-milking ewes with an abundant supply of milk provide their lambs with more milk than they can handle, and the lamb is found scouring severely (having diarrhea) or lying on the ground with the head drawn back and the legs "paddling." Lambs in feedlots may seem perfectly healthy one day and drop dead the next. The culprit is a germ known as *Clostridium perfingens* Type D, which is normally present in the bowels of most sheep. Under circumstances brought about by heavy feeding, the germ grows rapidly and produces a poison that is absorbed through the intestinal wall, with fatal results. Sometimes death occurs so quickly that the shepherd does not even notice anything wrong with the animal.

There is some evidence too that tapeworm infestation may increase the likelihood of enterotoxemia. A heavy load of tapeworms interferes with the movement of the food within the gut during the digestive process, and this interference may have the effect of converting protoxin to toxin and toxemia. Also, the feeding of the worms may affect the gut wall in such a way that the toxin moves from the gut into the bloodstream. In the ordinary course of events, tapeworms do not seem to have a harmful effect on sheep, either as far as weight gain or general condition, unless the infestation is so heavy as to cause intestinal blockage.

Losses in young lambs may be prevented by vaccinating the pregnant ewes with Clostroid D toxoid during the last two months of pregnancy, timing the second shot for at least two weeks before lambing. This not only protects the ewe but the effect carries the lamb through docking time with the additional immunization of tetanus antitoxin, which is incorporated in the vaccine. The vaccine is available from your veterinarian as Fort Dodge Vaccine Clostroid D. Doseage does not vary with nor depend on age or weight. Follow the label directions and, if the ewes were not vaccinated before lambing, vaccinate the lambs.

Symptoms of enterotoxemia are general sluggishness and diarrhea (symptoms that also indicate worm infestion). Once the lamb is down, with head thrown back, treatment is usually of no avail. General preventive measures include avoiding sudden changes of feed, introducing the lambs to a gradual build-up of grain feeding, feeding hay before the daily ration of grain, feeding frequently and regularly, and keeping lambs free from parasites. Bottle lambs should be fed little and often, six to eight ounces every four hours, skipping the 2 A.M. feed.

PNEUMONIA

Starvation and pneumonia of baby lambs are the biggest causes of mortality. A watchful shepherd, well-cared-for ewes, and a well-set-up barn are of much more use in overcoming these two killer conditions, which go hand in hand, than all the shots and medicine that you can round up.

The lamb comes into this inhospitable world soaking wet, sometimes into near freezing temperatures. If it is not promptly dried off and fed, its body temperature drops and its resistance is lowered to dampness, drafts, and the infection that is carried in the nasal secretions of the ewe or in her fleece. In damp or unseasonably warm humid weather the ewe's fleece becomes a soggy nest of bacteria, viruses, and parasite eggs. The newborn lamb, in its attempts to find the nipple in the forest of wool under the mother, gets a mouthful of this soiled wool, along with the germs and bacteria harbored there. Careful crutching will eliminate this condition.

A ewe that has a snuffly nose will spread infection on the lamb as she licks it and on the hay she eats and in the water troughs. The best medications to control this source of infection are the sulfonamides. They are inexpensive and can be mixed with the feed. Sulfamerazine or sulfamethazine powder is mixed at the rate of one pound per forty-five pounds of feed. The feed must be ground to keep the powder evenly mixed. There are various sulfa medications, such as Sulmet and Diamed, available from drug suppliers that will serve the same purpose. Medication through drinking water is not satisfactory because there is no way to control the sheep's intake.

PINNING

This is a condition in which the feces collect under the tail of the baby lambs, causing the tail to become stuck to the body and blocking the anus. The affected area should be cleaned by scraping away the feces. Rubbing cold cream or vaseline around the area under the tail and along the underside of the tail will prevent fecal material from sticking to the body.

WHITE MUSCLE DISEASE

This may occur on many different types of soils and pastures, but is most common in areas of good rainfall and among lambs grazing in legume pastures. The condition is due to a low level of selenium (or vitamin E deficiency). It has been found that alfalfa hay contains a substance that interferes with the body's proper utilization of vitamin E. Affected animals breathe abnormally and will stand with the back humped. Growth will be stunted, and they have a low resistance to infection. If dead lambs are autopsied, the muscles of the legs will be found to be chalky white.

To prevent the occurrence of this disease it is helpful to feed the pregnant sheep wheat in their rations, and when the lambs are born, to include linseed oil and beet pulp in their feed; these three feeds are high in vitamin E. Where a selenium deficiency is suspected, the injection of 0.25 cc. of injectable vitamin E will often give the lamb a rapid pickup. The vitamin-selenium compound for young lambs is L-Se, and for older lambs and sheep, Bo-Se. Bo-Se should never be used for young lambs, as the high selenium content would kill them. Both compounds are available from a veterinarian. The administration of vitamin E by mouth is not of much use.

STIFF LAMB (POLYARTHRITIS)

This is an infectious disease of young lambs. There is stiffness, loss of weight, and unwillingness to move. The lamb may be down on either the front or hind legs. The lamb's appetite remains good, but unless it is hand-fed it will starve because it cannot follow its mother to nurse or walk to the feed troughs. This disease and white muscle diseases are very similar and often confused, and in both cases vitamin E deficiency has been shown to be the cause. Injections of 3 cc. of penicillin daily for three days will be of benefit, in addition to the vitamin E, the dose of which depends on the brand of vitamin E-selenium compound that is used. Affected lambs should be put in a dry, clean pen and then bottle-fed at regular intervals. If the young lambs are confined with their mothers and assisted in nursing several times a day, the mother's milk supply will continue and the lamb can nurse again when both are returned to the flock.

ENTROPION, ECTROPION

Entropion is a condition of the eye of young lambs in which the lower eyelid turns in, causing irritation. This condition may cause blindness unless it is corrected. If the condition is not serious, daily application of opthalmic ointment will usually suffice. Should this do no good, it may be necessary to pick up a fold of skin sufficient to draw the edge of the eyelid to a normal position, and cut off the fold so that the eyelid will be drawn away from the eye as healing takes place. A small stitch will hold the skin in the right place.

Ectropion means that the lower lid turns out, making a trap for seeds and dust that irritate the eye. Use eye ointment under the lower lid, and in summer provide shade, darkness, and fly control.

WEAK LAMBS

When a lamb is discovered cold, weak, or apparently lifeless, the most important steps are to get it warm outside and inside. First, put a hot water bottle covered with a thick towel in a box, and put the lamb on top, covering it with more cloths. Put the box near a radiator or some other source of heat. Using a medium-size needle and a 10 cc. syringe, slowly inject 5 cc. of sterile glucose solution under the armpit of the lamb. Cover the hole with a finger when the needle is withdrawn to prevent the solution from leaking out, and in a few minutes repeat the process on the other side. Be sure that the needle is only just under the skin and not into the muscle, and depress the plunger slowly. Sheepmen Supply Company of Barboursville, Virginia 22923, supplies a commercial product for this purpose, manufactured by Anchor under the name of Caldex, which combines dextrose with calcium borogluconate and magnesium chloride. From 2 to 5 cc. of this will give the lamb's body something to feed on, producing energy and internal warmth. The injection of 2 cc. of penicillin will ward off pneumonia.

Keep the lamb in its nest until the mouth feels warm to the finger. As the body warms, the teeth will unclench and it will be possible to smear ¼ teaspoon of honey on the lower teeth, using a popsicle stick or tongue depresser. Don't try to give anything else by mouth until the lamb is up and bleating because it will be unable to swallow and the liquid will go into the lungs, causing mechanical pneumonia.

When the lamb is able to run around and suck vigorously on a bottle it can be returned to the barn. Still, it will need to have access to a source of warmth, such as a heat lamp, and also regular supplementary feeding if the mother's milk supply is poor. Lack of milk, or milk of poor quality, is the usual reason for a lamb getting chilled. If the dam is a consistently poor milker she should be culled, but it may be a case of a bigger lamb grabbing all the available supply.

When making inspection tours of the barn, always investigate the lamb that is off in a corner sleeping. If, when you put a finger in its mouth the mouth feels warm and the lamb gets up and runs to its mother, all is well; but if the mouth is cool and the lamb seems listless, get some warm milk into it before matters become more serious.

For a cold and weak lamb a heat lamp can be a lifesaver, and when properly used is one of the most useful tools at lambing time. It is also potentially dangerous if *not* properly used. Lamps must be suspended by chains or thick cords, not by the electric cord, and must be protected by wire guards round the bulb. Lamps with these guards are sold by poultry suppliers. The socket for the lamp must be ceramic, not plastic, and the cord must be heavy duty and rubber insulated. Use only underwriter-approved lamps. The light should be at least thirty inches from the level of the bedding and should not be used where the ewes might bump against it. To gain the best benefit from a heat lamp, section off a stall in the barn, put in a trough of grain and a pail of water, and hang the lamp where the lambs can comfortably gather underneath it. A lamb creep panel or gate panel with wide-spaced bars will keep the ewes out without preventing the lambs from going back and forth to their dams as they wish. When the weather gets warmer, turn off the lamp so that the lambs do not become too conditioned to heat.

ORPHAN LAMBS

Every lamb should have a mother and every ewe should raise a lamb—and if lambs and ewes could read they would probably agree—but some ewes who have lost their lambs firmly resist all attempts to make them adopt an orphan. Of the various methods that can be tried, none is surefire, but patience is a big help. Remember, a lamb is better with a sheep for a mother than it ever will be with a bottle.

Suppose you have a mother who for some reason cannot care for her lamb from the time of birth, or whose mother died. If you should be so lucky as to find another ewe that is in process of lambing and which delivers only one lamb, take the orphan, wash it in warm water and then quietly take as much as you can find of the birth fluid and afterbirth from the second ewe and smear it all over the orphan. Place this second lamb under the new mother's own lamb, and with any luck she will take your word for it that she has borne two lambs, and will take care of both of them.

If a newly delivered ewe refuses to take her lamb, either because of a difficult delivery or fright and inexperience, pen her up in a narrow stall with some warm water and good hay. Fix a barrier across the stall so that the lamb can be near her but can get away if she is too rough with it. Then, when the ewe is calm and rested, make sure the teats are not plugged and hold the lamb under the mother, putting the teat in its mouth if it is slow to start nursing. You may have to do this several times before the ewe will accept the lamb, but it nearly always works. This same method can be used to graft a rejected twin onto a ewe whose lamb has died, even if the twin is one or two days old. Some ewes do not mind having a strange lamb in the pen with them, and these can be left loose in the stall and just steadied with the hand at feeding time until they get used to the lambs' nursing.

There are other methods of securing adoption. One is to skin the dead lamb and put the skin over the orphan, so that the mother thinks the orphan is her own lamb because the smell is the same. No doubt this works, because it is often recommended, but it is a mess, and not many ewes are deceived so easily. Sometimes it helps to put camphorated oil or Vicks on the ewe's nose, and around the tail and on the shoulders of the lamb. "Mother-up" spray is sold by livestock supply houses for the same purpose.

An orphan who made good.

If none of these methods suffice, then the lamb must be raised on a bottle. Raising orphan lambs is not difficult if a few simple rules are followed. First, ewe's milk is very rich in fat, much richer than cow's milk, and therefore if the lamb is to be fed cow's milk, it should get the richest milk possible— the creamy kind that Guernseys give. Goat milk is ideal. If a daily supply is not available, it can be purchased in bulk and stored in the freezer. If an artificial milk-replacer is to be used, make sure it is one specially formulated for lambs, containing at least 30 percent fat. Land O'Lakes, Ewelac, Albers Lama, and Nursette are suitable. Never try to raise a lamb on dairy calf milk replacer or Carnation nonfat dried milk, because the lamb will become potbellied and stunted and will probably die. Feed only a few ounces at a time at first: the flanks should be level with the ribs, not bulging out. It is much better to give small feedings of rich milk and to have water available for the lambs than to be overgenerous with a lower-fat and possibly cheaper substitute. If the lamb shows signs of scouring, cut out the milk feeding for a few hours and give warm water to which a teaspoon of limewater (from the drugstore) has been added.

If several lambs are to be artificially reared, it might be

Milk bottles with rack.

worth investing in a self-feeder. For baby lambs the bottles and racks shown here are satisfactory. In these the milk is contained by a special lamb nipple. The feeder should be placed on a bed of shavings or sawdust that can be cleaned frequently, because the nipples tend to drip. As soon as the lambs are about a week old and able to suck more strongly, they can be introduced to a feeder that incorporates a tank or pail with tubes leading from the milk to nipples placed in a framework set higher than the liquid, so that the milk must be drawn up by the lamb (see illustration). Milk is fed warm from the bottles, but the milk in the tank can be cold: the lambs will like it and will be less apt to make themselves scour by gobbling too much at one time. Sheepmen Supply sells a Lambar that serves this purpose very efficiently. Make sure that the milk is always fresh and sweet and that the tank or bottles are kept scrupulously clean.

Homemade milk feeder. The nipples are set higher than the liquid so that the milk must be drawn up by the lamb.

Most of the troubles people have in rearing bottle lambs arise from improper feeding. Lambs cannot drink eight or ten ounces twice a day, but must be fed every four hours, six to eight ounces at a time. They can go through the night without a 2 A.M. bottle, provided they are warm and comfortable. Bedding must be dry and clean, and area selected for their pen should be light and well ventilated; a heat lamp, suspended three feet above the floor, will increase their comfort when the weather is very cold. Lambs are playful, and will enjoy having a bale of straw or hay to jump around on, as well as room enough to exercise.

It is better to keep orphan lambs in the barn than in the kitchen, where they are very appealing the first day and thereafter a mess. Lambs kept in the barn with other lambs will start to nibble grain and hay at an earlier age than if they had no sheep to copy. By the time they are four weeks old they should be well started on these feeds, and bottle feeding can be gradually decreased. If the lambs are using a Lambar or other self-feeder, make the dry feeds available near the milker and place a pan of clean water near the grain trough.

Lambar, commercial milk feeder.

TENDING EWES

Ewes that are nursing young lambs should receive from one to 1½ pounds of grain daily—two pounds if they have twins—and as much good hay as they will eat. At this time, very heavy demands are made on the ewe's body to produce liberal quantities of milk that will ensure strong, healthy lambs. A lamb does not start eating solid food until it is three weeks old, and the rate of growth during this period therefore depends on the quality and sufficiency of the dam's milk.

Milk production often increases when succulent feed such as silage is given. Silage, plus good hay and a daily grain ration, will fulfill their needs until they return to pasture. Those ewes that are nursing twins should, if possible, be separated from those raising singles, so that they may receive higher rations of concentrates. Plenty of clean water with the chill taken off should be available at all times.

When pasture season arrives, the feeding of silage and concentrates may be discontinued—gently, to avoid an abrupt change of feed. It is best, however, to feed some hay every morning until the ewes and lambs adjust to pasture. If the ewes are free from worms and parasites, they should be able to produce plenty of milk on good pasture until the lambs are weaned.

EAR-TAGGING AND PAINT BRANDING

Ear-tagging serves a variety of useful purposes. It is, of course, one sure way of identifying each sheep and is essential if proper records are to be kept. The use of different-colored ear tags makes it possible to mark each sheep so that one can tell at a glance how old she is, or, if the flock is divided into groups at breeding time, to which group each sheep belongs. Breeding and performance records, wool production and lamb production, and mothering qualities can all be much more accurately recorded when each sheep is properly identified.

At lambing time each lamb should be marked with a number and its dam marked with a corresponding number. Codes can be used; for instance, all the lambs born within a seven-day period could be marked one color, and the color changed for the next seven-day period, and so on, so that at the end of the lambing season one could tell at a glance how old each group of lambs were. Lambs that are twins could be marked differently from singles and ewe lambs could be marked differently from ram lambs. The chief advantage of this color marking is being able to match up mother and young. When a ewe is seen in the pasture, or a lamb is found alone and calling for its mother or is seen to be sick or injured, knowing which lamb belongs to which ewe makes it possible to be sure that every sheep and lamb is present and accounted for. Otherwise, there might be lost, strayed, or sick animals out in the pasture. The marking is done with washable water-based wool paint, available from a livestock supply store, and is applied by using branding numbers, supplied by the same source.

Tattooing is not a good substitute for ear tagging. The numbers become faint after a few years and are hard to read at the best of times.

TENDING RAMS

Rams should be penned separately from lambing ewes to avoid risk of injury to ewes and lambs when the rams push their way to the feed racks. A corner of the barn or another building will do, but wherever they are put, the rams should not be neglected. They need good food to regain the condition they lost when chasing up and down the pasture after the ewes.

When the ewes go out to pasture the rams can go with them, as the latter are not usually aggressive at this time and can be safely grazed in the same field as the lambs. New rams, if required, are introduced to the flock in summer.

Some caution is advised before bringing an additional or re-placement ram, as the incumbents may fight the newcomer and, if one ram is put in among a number of rams who are used to each other, these may gang up on the new ram and do him serious injury. Pen the new ram up with the others for a few days in very close quarters, so that there is no room for any of them to back off and charge the others. Usually within a week they will be heartily sick of each other and will go off quietly into the flock, but watch them at first until you are sure all is well, and never leave them alone just to fight it out. The bleeding heads will be an invitation to flies. Once they are getting along well together, the rams will tend to go around as a group during the summer months and will seem to like each other's company. But some rams just never get used to each other and will always fight. In this case, they will have to be kept separated.

Ram lambs can stay out on the pasture with their dams until the time comes for them to be brought into the barn for the supplemental feeding that gives them a finish before but-chering time. Small horns may grow on both ewe and ram lambs that have horned breeds in their recent ancestry. These horns are thinner and lighter than the true horns and may break off. Usually the spot will heal over, although a sprinkling of Bloodstopper will help to keep flies away from the sore spot. It's a little disconcerting to grab that handy handle to move the sheep and to have it come off in one's hand, but this does not hurt the sheep. If any bits of skin and horn are left hanging, trim them off. A real horn that is broken is a more serious matter, for the horn is part of the skull and does not shed as an antler would. The wound must be carefully cleaned and covered with an antiseptic dressing. The dressing is to be kept on and replaced when necessary until the wound is completely healed, which may take two or three months.

5

Feeds
and
Pasture
Management

In order to provide adequate pasture and feed for sheep it is necessary to know something of the composition of these feeds and how the sheep digest them.

THE DIGESTIVE PROCESS

Before the body can use food for growth or maintenance, that food must be digested and made soluble. The process of digestion begins in the mouth where the food is moistened by saliva, an alkaline fluid that contains a small amount of an enzyme which changes starch into sugar and thus makes it available for use by the body: sugar is soluble, but starch is not.

When the food has been chewed just enough to be easily swallowed, it passes into the first stomach, the paunch (or rumen), an organ with rough walls that holds about three gallons. The food is ground against these rough walls to prepare it for further digestion, and is acted upon by certain bacteria. Some of these bacteria can be destroyed by indiscriminate medication, such as excessive use of antibiotics. When sheep or lambs suffer a breakdown in the digestive processes, in severe diarrhea for example, dosing with acidophilus milk or small feedings of yoghurt will often start the stomach bacteria working again.

From the rumen the food passes into the reticulum and then the omasum, the second and third stomachs, where further bacterial action works to digest the feed. This bacterial action produces some warmth and is one of the reasons why ruminants are able to digest roughage.

While this digestion is going on considerable gas is formed, causing the sheep to belch contentedly while they are enjoying rehashing the day's meals. When gas forms too rapidly and there is no belching, the sheep will bloat. Also, forage that is too constipating, lack of water, or occasionally the weight of an unborn lamb may cause impaction of the rumen or omasum. (Chapter 6 discusses ways of avoiding and alleviating these conditions.)

When the sheeps' hunger has been satisfied, they will choose a quiet place and settle down to chew the cud. This cud is a small wad of food that is returned to the mouth for a thorough chewing. Sheep will chew about seven hours out of twenty-four and need a quiet place where they can do so in peace. They will not chew when frightened, in pain, or harassed; and if they don't chew, the food ferments in the stomach, causing colic and bloat.

The second stomach acts mostly as a passageway for the food. In a cow this stomach is the place where "hardware" lodges, but sheep are such careful eaters that this problem seldom arises. The third stomach receives the food from the reticulum by the action of the esophageal groove. This latter acts rather like a railroad switch, shunting the food in whichever direction it is supposed to go. This third stomach, the omasum, or "many-plies," is full of leaves that rub together to grind the food to a fine paste before it passes into the fourth stomach, or abomasum. In this last phase, the food is acted on by the digestive juices, including pepsin, rennin, and hydrochloric acid. The fourth stomach starts the digestion of the proteins.

So far, very little of the food has been absorbed. This takes place in the small intestine where juices and enzymes from the mucous lining of the bowel do their work. This small intestine is about eighty feet long, and in it most of the dissolved materials are absorbed through the walls and into the blood. The whole digestive process takes between three and five days.

PROPER FEEDING

The business of feeding sheep depends to a certain extent on the nutritive ratio of what is fed, on the amounts of total digestible nutrients, and on the composition of the feeds given. There are many books dealing with sheep and farm animal nutrition that give detailed and explicit tables of methods of balancing feeds. Such information can profitably be used by those growing all their own feeds, and by owners of large flocks whose feeds are mixed in very large quantities. Sample rations are given later in this chapter. The composition of these will vary somewhat with locality and season, but in general will provide adequate feed for the farm flock.

Well-fed sheep should look healthy and blooming, and be active and clean-looking. In addition to the eye, the hand of the shepherd can measure the condition of the animals by pressing down on the back gently with the fingertips to feel for broad, sturdy backs or spines sticking up like a ridge.

While sheep do, indeed, make use of a great variety of grasses and plants, they are not omnivorous. I have seen a flock of Cheviots half starved in a barn where the hay racks were full and the feed troughs loaded with grain. The trouble was that the hay was stemmy timothy and the grain was dry corn chops. After a few weeks on mixed second-cutting hay and a simple mixed grain ration the difference was like night and day. Overfeeding is just as bad. A ewe who is roly-poly fat may not breed, and if she does she is likely to have trouble at lambing time. She also runs the risk of "backing," or getting over on her back and being unable to right herself. In Shetland, it is the obligation of anyone seeing a backed sheep to take steps to right it, because sheep cannot breathe properly on their backs and will suffocate if not helped. Even the school bus will stop if the children see a sheep in this predicament. A further danger is that overly rich feed can cause scours, bloat, or even death.

PLANNING RATIONS

Estimating the amount of feed required per sheep per year depends on a variety of factors—length of winter season, size of ewes, and the kind of hay and grain being fed. As a rough guide, allow five hundred pounds of good hay to carry a ewe from mid-October to mid-March. Since grain is usually purchased once every two weeks or once a month, the quantities can be adjusted according to the ewes' needs, allowing one pound per ewe per day average for thirty days before lambing and 1½ pounds per ewe per day for thirty days after lambing.

Unless you know from which field the hay was cut, ask to have one bale broken open and inspect it before buying any large quantity.

WATER

Water is not usually considered to be part of the feed, but the provision of good, clean water is vital if the sheep are to be able to derive nourishment from their feeds. Feeds with a high moisture content help the digestive system function properly, but dry feeds require plenty of water in the paunch if they are to give full benefit. Food is moistened thoroughly in the rumen before being returned to be chewed as cud, and plenty of water is needed for the proper supply of saliva to make the cud moist. In winter, the water should be kept above freezing either by the use of heating apparatus or, if large tubs are used, by the addition of sugars to lower the freezing point of the water. The best product to use for this purpose is molasses; it contains dextrose, levulose, sucrose, and minerals, as well as being palatable to the sheep.

CARBOHYDRATES

Starch, sugar, and fiber are found in grains such as corn, oats, and barley. Fiber is found in stems, such as hay, and in the hulls of grains. Carbohydrates provide energy to keep the animal warm, to keep the body processes working, and to make fat.

FATS

All feeds contain some oils. These work in the same way as carbohydrates, but are more concentrated.

PROTEIN

Protein contains carbon, hydrogen, oxygen, sulfur, phosphorus, and nitrogen, and is an essential component of any feed. It is needed to ensure growth, milk production, replacement of body tissue, and the growth of wool. Because wool is almost entirely pure protein, sheep require more of this substance than cattle do, particularly during pregnancy when the body is not only nourishing the fetus but is continuing to grow wool. Legume hay, clover, alfalfa, and soybean are good sources of protein, and if these are in short supply the ration can be supplemented with linseed meal, cottonseed meal, or soybean meal.

MINERALS

All animals need minerals in their food in order to live. Even a ration well supplied with the necessary nutrients will not sustain life if minerals are lacking. As examples, the nuclei of the cells of the body are rich in phosphorus, the skeleton is largely composed of phosphorus and calcium, and cobalt is important in aiding the rumen bacteria in manufacturing vitamin B_{12}, which is important for growth.

The U.S. Plant, Soil, and Nutrition Laboratory at Ithaca, New York 14850, has published a series of maps showing where mineral deficiencies occur, and mineral supplements can be fed to offset these deficiencies. Poor wool is often a sign of mineral deficiency, and a chemical analysis of the wool will pinpoint the particular mineral which is lacking.

Minerals can be added to the grain ration when it is used, and at other times should be offered free-choice alongside the salt. Some feed mills sell premixes containing the necessary minerals plus vitamins. These minerals will mean a higher percentage of live lambs, a better supply of milk for the ewes, an increase in the wool clip, and faster growing lambs.

SALT

Salt is necessary to all grazing animals. It should be fed free-choice and can be offered in a variety of forms. T.M. Salt Bricks are a brand of block to which minerals have been added, and are adequate in most areas. There is some objection to bricks on the grounds that sheep wear down their teeth by chewing on them.

Sheep will eat as much as twelve pounds per head per year of salt, considerably more than cattle do. When sheep are deprived of salt they develop a voracious craving for it; if sheep unaccustomed to salt are allowed sudden free access to it they may eat too much and poison themselves. Until animals are used to it, salt should be hand-fed, starting with very small amounts.

Because sheep like salt so much and eat it so freely, it is a good vehicle for other minerals and for low-level wormers; you can buy treated salt and salt blocks, or make up your own salt-mineral mixture.

WINTER FEEDING

In areas where the pasture is dormant during the winter, the following are good winter rations (in pounds). For in-lamb ewes, start at one pound a day, increasing to one-and-one-half pounds immediately after lambing (two pounds for ewes with two lambs).

1. Ground ear corn, or shelled corn   60
    Whole oats   20
    Wheat bran   10
    Free-choice legume hay

or,

2. Corn   300
    Oats   200
    Wheat bran   100
    Soybean meal   100
    Molasses   180
    Minerals   8
    Salt   15
    Bone meal   24
    Beet pulp   100
    10 bales good hay

or,

3. Mixed dairy feed containing 12 or 14 percent protein, and good hay.

or,

4. Grass hay and mixed dairy feed containing 18 percent protein.

Ration 2 makes about ¾ ton of feed in the quantities given. It has the advantage of containing sufficient roughage that it will not cause the sheep to bloat if they should accidentally overeat. It is also an adequate and palatable ration for young lambs.

Additional free-choice salt and minerals should also be offered with the above rations.

Root crops such as turnips and carrots will keep ewes healthy when added to the feed, as will chopped cabbage and pumpkins. The pumpkins carry a bonus in that the seeds have anthelmintic properties, helping to keep down worms.

HAY

The daily ration of hay should be clean, green, and not too coarse. Timothy is not suitable hay for sheep unless mixed with other hays, as it is too stemmy and inclined to be constipating. If you have to buy hay, clover or alfalfa are more expensive, but the sheep will discard less and do better and in the end save you money. Mixed clover, alfalfa, and timothy are satisfactory, as are mixed clover and grass hays, but care must be taken to see that they are not mostly mixed weed. Orchard grass is not suitable unless cut very young.

Legume hays are rich in protein, calcium, and vitamins A and D, but if this is not available mixed grass and legume hays will serve quite well. When choosing hay, look for green, tender hay, with fine stems and leaves. Discard any moldy hay because it is dangerous feed, and avoid old brown hay, as it has little food value.

SILAGE

Silage consists of green plants, harvested early and chopped, compressed, and allowed to ferment. During fermentation, part of the plant sugars are broken down, organic acids are formed, and carbon dioxide gas is released. Silage is easy to store and feed, it is free of dust and chaff, and the sheep enjoy it.

Satisfactory silage for sheep can be made from a wide variety of plants, including cereal grains, grasses, pea vines, and beet tops; corn, sorghum, and soybeans combined will make a good silage that can be fed in conjunction with legume hay as the roughage ration. If legume hay is not available, some other protein supplement must be provided. The ration of silage is usually limited to about three pounds a day, with the rest of the roughage being provided by hay.

If no silo is available, the silage can be stored in a trench or in a pile on the ground (provided there is adequate drainage for the juices produced by the material) and fed through the winter. The sheep must not be allowed to drink the liquid seeping from the silage, nor to eat the grass in the immediate area where it drains.

PASTURE

Allow the sheep to run on pasture as long as possible. When the fall grass is watery and frost-nipped, hay— preferably a good legume—can be put out each day and they will eat what they need. For a supplemental fall pasture, Piper Sudan grass, seeded in May at the rate of thirty pounds per acre, will be ready to graze about six weeks after planting in the East Central States. One acre will support four animal units, or twenty to twenty-five ewes. This is a good hay for very dry seasons.

Sheep will also enjoy grazing corn stubble, but should be given some other feed before being turned into the cornfield in the morning, and should not be allowed to graze more than a few hours.

Many of the herbs that sheep find on natural grazing land are good for them. Wild raspberry leaves are particularly good for sheep during pregnancy, strengthening the pelvic muscles and contributing a tonic effect. Juliette de Baïracli Levy, in her book *Herbal Handbook for Farm and Stable* (Rodale Press, Emmaus, Pennsylvania, 1976), lists the uses and preparation of a number of herbs of value to sheep and gives direction for making drenches and brews with herbs.

Wilted and fallen wild cherry leaves can be poisonous to sheep, but usually the animals will not eat them unless they are very hungry. And wet clover or alfalfa will cause bloat.

There is another source of poison for which one must be on the alert. Trash dumped or thrown from passing cars sometimes contains such lethal substances as lead paint, poisonous cleaning agents, and old medicines which the sheep, being inquisitive creatures, may lick. Paint chips or splashes from overhead powerline towers can be fatal; when towers are being painted, keep the sheep away until all trace of paint can be removed from the ground below.

During the peaceful months of late autumn and early winter the ewes should go out every day and the shepherd should see that they get enough exercise. If they are given hay in the morning and are then free to roam, they will usually stay on their feet most of the day, nibbling here and there and coming back to the barn or loafing pen in the early evening. But if they are kept in fenced fields or pens, some system must be devised to ensure that they get enough exercise. One way is to feed the hay at some distance from the barn, or to take them for a gentle walk of a mile or so with the assistance of a quiet, well-trained dog. This is also very good for the shepherd. The dog must be very quiet and obedient, as an excitable barking dog is a menace among sheep and should never be allowed near them.

SUMMER PASTURE

Summer pasture is the natural environment for sheep. They will thrive on the various grasses, legumes, and herbs they find, and will browse among woody plants and bushes that grow wild. In doing so, they are converting into usable protein, in the form of meat, a great deal of grass and wild growth that would otherwise not be used. If some cattle are grazed with the sheep, the former will eat the larger, coarser herbage that sheep do not like, while the sheep will eat the tender young stuff that the cattle ignore.

There is little truth to the assertion of the Old West cattle-men that sheep kill the grass. They do this only when so starved that they have to eat the pasture right down to the ground. On the contrary, their habit of wandering around to graze and their pelleted droppings build up the soil. Because they have such an appetite for tender young growth, however, sheep should not be put on the pasture too early in the spring, or the new growth will not catch up before the hot summer weather.

Stocking rate for pasture depends on the size of the flock and the productivity of the pasture. It is customary to count seven mature sheep or fourteen lambs as one animal unit and one cow as an animal unit when estimating acreage required. The following description of pastures and forages will help you determine the needs for the flock.

PERMANENT PASTURE, UNIMPROVED

This consists mainly of bluegrass and weeds, and rates fair to good in spring and fall, slowing down in summer. Allow about two acres per animal unit, plus an extra half acre during midsummer. Lime and fertilizer can increase the productivity of this kind of pasture.

PERMANENT PASTURE, IMPROVED

These pastures may consist of bluegrass and white clover, timothy and clover, birdsfoot trefoil mixed with timothy, and clover or alfalfa-grass mixtures. Liming and fertilizing should be carried out according to soil test, and the pastures will provide excellent grazing through the summer. One acre per animal unit, grazed rotationally, will sustain the flock throughout the pasture season, but the land should not be used during the early spring and late fall. Permanent bluegrass or fall-seeded rye can be used in spring, and permanent bluegrass or Sudan grass will provide late fall pasture.

Many homesteads and small farms have land that is growing a variety of grasses and weeds that will, when mowed, prove very appetizing to sheep. Mowing improves any pasture by removing dead growth and keeping down thistles and clumps of inedible weeds. The more it is grazed, the more grasses will thrive on rough land, causing the weeds to die out, but such meadows will not support so many sheep as will permanent improved pasture.

When a small flock is turned into pasture, the available land should be divided into at least two and preferably three separate areas. The sheep are grazed in each area for three or four weeks at a time and then moved to the next area, thus allowing the crop to grow back and leaving worm larvae to die out on the unused pasture. Even though the pasture may be eaten down hard, long ripe grasses or thick clumps of uneaten brush and weeds should be clipped to allow the better grasses and legumes to grow and admit the sunlight to get to all the plants and kill off worm larvae.

WOODS AND SWAMP

Woodland and scrubland will not support sheep without supplementary feed. The sheep will clear blackberries and wild raspberries, poison ivy and other wild growth, but must have access to better pasture or hay if their breeding and mothering abilities are not to suffer. Low, swampy ground is not pasture—it is an invitation to foot rot, liver fluke, and a multiplicity of other ills.

ALFALFA AND CLOVER

Alfalfa is one of the best pastures, and if allowed to rest between periods of grazing it will stand pasturing for many years. Legumes such as clover and alfalfa are not only nutritious but they benefit the soil through the addition of nitrogen as well. The seed is expensive and requires some care if a good stand is to be established. The highest food value is contained in the fully matured plant, but younger and less-succulent plants are less likely to cause bloat. Alfalfa fields can be used for both grazing and hay crop in the same year.

Clover for sheep is usually used in conjunction with timothy or grass or birdsfoot trefoil. Pastures sown with clover alone may cause bloat. Both clover and alfalfa can cause bloat in sheep when the flock is allowed to graze wet fields, particularly if they go out early in the morning when they are hungry. When sheep have been allowed on these pastures they must be brought home very slowly because chasing by dogs to bring them home at a run, particularly in hot weather, may cause some of them to collapse and die.

If sheep find a hole in the fence, they will usually take advantage of it, but they do not travel long distances, as steers are apt to do. Sheep can be safely let out to graze on open land as long as there is no danger of their getting to something forbidden or on the road. They can be easily brought home at the end of the grazing period as long as they have been around the same place long enough to know where home is.

6

Keeping Sheep Healthy

Healthy sheep are a pleasure to manage. They return greater profits than sickly ones ever do, the ewes are better milkers and live longer, and healthy sheep produce healthy lambs which, in turn, are more profitable.

GENERAL PRECAUTIONS

There are some general measures that should be followed to prevent diseases and parasites in the flock, and some precautions that will help avoid accidents and assure the flock's well-being:

1. Provide clean grazing by rotation of fields and pastures.
2. Worm and dip all sheep at least once a year.
3. Keep barns and buildings clean and well-aired.
4. Keep newly bought sheep separate from the flock until you are sure the newcomers are not harboring disease or parasites.
5. Destroy dead sheep, either by burning or burying in lime. Never allow dogs to eat sheep carcases. Cysti-

cercus (or hydatid tapeworm infection) in sheep and dogs is the result of dogs passing tapeworm segments in their droppings on the pasture, which are then picked up by the sheep. The eggs develop into cysts in the sheep muscle (heart, liver, lungs, etc.). A dog eating the uncooked or unfrozen carcase of a sheep may be eating cysts which will in turn develop into tapeworms, and so the cycle continues.

6. Feed generously and regularly, providing salt and clean water at all times.

7. Look over the sheep at least once a day to detect signs of sickness or trouble.

8. Never put out hay with the string still on the bale or leave the looped twine lying on the ground. There is danger of the sheep getting tangled with the twine round the neck and feet.

9. Push down the handles of all water and feed buckets. The sheep may put their heads deep inside the buckets and get the handle caught around the neck, and it's no easy task to catch a sheep that is running frantically round a field with its head in a bucket.

10. If gates or panels are left in pen or pasture, make sure they cannot fall on young lambs and injure them.

11. Feed hay and grain in racks so that the sheep do not walk around in it. Arrange waterers so that they are not fouled by droppings, and see to it that the surrounding area is well drained.

12. Keep equipment clean and sterilize all needles and docking tools.

13. Keep barns and equipment in good repair. Mend jagged pieces of fencing and hammer down protruding nails.

14. Never combine two different medicines or exceed recommended dosages.

15. Nature has given sheep beautiful wool coats to keep them warm, and nimble feet so that they may travel far for their grazing, but she has certainly shortchanged them in the matter of self-defense. Even the

impressive horns carried by the Dorsets and Rambouillets are rarely used as weapons. Only when her lamb is threatened will a ewe become aggressive, and even then not all of them do so. A protective mother will use her head as a battering ram and is capable of routing a large dog. Because of their lack of means of defense, it is the responsibility of the shepherd to see to their safety: to provide adequate fencing to keep out predators and to ensure that the farm dogs are not allowed to worry the sheep.

WOUNDS

Because they graze quietly, sheep seldom suffer wounds and lacerations during a normal day. Simple scratches from barbwire fences or sharp branches will usually heal quickly of their own accord, but when noticed, should be sprayed with scarlet oil or dusted with Bloodstopper (haemostatic powder) during the summer months to keep flies from gathering on the wound and laying eggs, which will hatch into maggots. More serious wounds, such as dog bites, deep shearing cuts, and lacerations from falls, should be cleaned and treated with antiseptic, then protected with an application of Bloodstopper or Vet Paint. If pieces of tattered skin or tissue surround the wound, snip them off with a clean pair of scissors before the wound is dressed. Sheep are very resistant to germ infections of the pus type, but if a wound is not clean and open, use antibiotics such as Combiotic or penicillin at the rate of 3 cc. twice a day.

Antibiotics are normally injected into the muscle of the side of the neck, as illustrated. Use a short needle, either 16 or 18 gauge, and sterilize it in alcohol between each sheep. Head the sheep into a narrow stall—the less room for it to turn around the better—part the wool at the site of the injection, and dab the skin with alcohol. Push the needle in about ¼ inch.

If the wound heals and is clean, more than two injections should not be necessary, but if the wound smells foul and pus gathers, meaning that infection has set in, the penicillin treatment will have to be continued for several days. In this case, give the injections on alternate sides of the neck. A subcutaneous (just under the skin) injection of tetanus anti-toxin is also a wise precaution, given at the rate of 1 cc. per 100 pounds body weight. Tetanus antitoxin gives immediate protection. Tetanus toxoid, however, requires three to four weeks to give the peak level of protection for one year; it is used in areas where tetanus has been a problem, and where protection against normal scratches and wounds is thought to be necessary.

It is very important to keep wounds clean. A simple washing with antiseptic solution or saltwater and the application of scarlet oil or Bloodstopper takes only a few minutes, but the treatment of a wound that has become infested with maggots takes much longer. The sheep suffers loss of condition both from the irritation of the maggots eating into the flesh

Preferred site for injections.

and from the fever resulting from infection. While it is true that the maggots clean the wound by eating away the infected flesh, it is also a fact that they do not stop there and will continue eating and multiplying, increasing the area of the wound.

To get maggots out of a wound, flush the area with several pints of antiseptic solution or saltwater. It may be necessary to pick out some of the maggots with tweezers. A liberal dusting of ordinary flour or cornstarch will choke them and make them crawl out; or, it is said, the touch of a copper coin will make them evacuate the wound.

Very extensive wounds, such as those suffered by sheep when attacked by dogs, are difficult to deal with, not only because of the problem of covering the exposed tissue but because of the effects of fright and shock. A dose of tetanus antitoxin, plus a thorough cleaning of the wounds, are important. Just as vital are rest, quiet, nourishing rations, and good nursing care.

A sheep in shock will stand with head drooping and ears cold, and will refuse to eat or chew the cud. In severe cases this condition may last for days and may even be fatal. Avoid overtreating, so as not to further frighten the animal, and supply lots of T.L.C. Penning in a familiar stable helps reduce the effects of pain and fright. Frequent redressing of the wound is unnecessary, and you needn't dose with antibiotics as long as the wound appears clean, although the torn tissue will smell foul as it dies and sloughs off. A veterinarian would usually cut off the torn tissue if called to treat such a wound, but sometimes the treatment further distresses the animal; the sheep will have just as much chance of recovery if left alone under the cleanest and quietest possible circumstances. As long as it will eat a little and drink and is not feverish, nature will get on with the job of healing.

DISEASES

A competent shepherd knows what constitutes the normal behavior patterns of his sheep and is on the watch for signs and symptoms denoting trouble. It is all-important to realize the importance of treatment before disease goes through the whole flock and before the animal or animals are so weakened that treatment is of no use. Some veterinarians are loathe to treat sheep and feel that sheep "die easy," possibly because some flock owners feel that a sheep is not worth the expense of the vet, and don't call him until the animal is on its last legs. It is no good waiting for days or weeks, trying this and that remedy to see if the sheep gets better, and then calling for help when the animal is indeed beyond help. Early diagnosis can save time, trouble, and money; with his expert knowledge the veterinarian can make his diagnosis and not only will the affected sheep still be in good enough condition to respond to medication, but the chance of an epidemic in the flock is greatly lessened.

It is sometimes difficult for an inexperienced shepherd to know when to seek help: a sheep with a wound, with a lame foot, or with a bad cough has an easily recognized symptom that can be described and treated, but a sheep that just stands around looking peaked and not getting any better is suffering from a more challenging disease, and it's best to call the veterinarian. Be on hand when he comes so that you can answer his questions, and then follow his recommendations. With the rapid advances made in treating livestock disease and maintaining livestock health that have developed during the last decade, many conditions formerly regarded as untreatable will now respond to modern medication.

Veterinarians are busy people, and in many cases it is possible to load the sheep into the back of a station wagon or pickup and take it into the office. This saves time for the vet, who has all his supplies and equipment at hand, and thus enables him to treat many more animals in a day.

To lift a sheep into the back of a pickup, clasp your right hand round your left wrist, clasp your assistant's right wrist with your left hand, under the sheep's middle, and have him take your right wrist in his left hand, making a sort of fireman's lift for the sheep. This enables you to lift the animal with a minimum of strain on both people and sheep; it is possible for two women or even two older children to lift a fair-sized sheep in this way. If you are by yourself, it is better to make a walk-up with an old door, using two gate panels as sides.

Important infectious diseases of sheep, with a few notable exceptions such as foot rot, are not common among farm flocks. Such diseases as anthrax, blackleg, tuberculosis, and rabies are described in books on animal diseases, but will probably never concern the average flock owner.

The first symptoms are inflammation, fever, and pain. When a body or part of a body is attacked by germs, the blood produces more white corpuscles to destroy the germs. Inflammation is a reaction of the body tissues to invading germs; pus is a mixture of dead bacteria, blood corpuscles, and serum fluids. A serum injected into the body tends to produce more white corpuscles to defend the parts attacked. Sulfa drugs absorb the food the germs live on, thus reducing their power to harm the body or to spread. An antibiotic destroys bacteria without affecting the body tissues. Antibiotics are effective only in cases of bacterial infection and are useless against viruses.

When sheep show signs of disease, every effort should made to treat the problem early because sheep tend to stay close together, thereby making the spread of disease more rapid. Sick animals should be isolated to keep the others from catching the disease and to permit the sick ones to receive special attention.

The normal body temperature of a sheep is 103°F. to 105°F. Special veterinary rectal thermometers are available, equipped with a loop on the end to which a string is attached, thus preventing the thermometer from getting lost inside the sheep. Cold ears are a reliable indicator that something is wrong, even though there may be no other signs.

Symptoms of diseases are so similar in many cases that no treatment should be undertaken just because it looks like something in the book. Get help from an expert and keep your troubles small.

ABORTION

This may be caused by eating frozen forage, by injuries, fright, or malnutriton. During the last six weeks of pregnancy, ewes should be protected from buffeting or worrying by other livestock or people, and care should be taken to see that they are not driven through narrow gates or doorways. Some cases are due to an infective agent, such as vibrio (infectious abortion), thought to be found in contaminated water, but this is not a common problem in farm flocks.

AFTERBIRTH, RETAINED

Occasionally a ewe will fail to "clean" after giving birth to a lamb. Various remedies are known. You can give her a draft of her own milk or a pint of water in which a handful of flour is mixed. Or, try raspberry leaf tea; make a strong brew of 2 handfuls of leaves to 1 pint of water, adding 2 tablespoonfuls of honey, and give a cupful frequently.

Because the ewe tightens up immediately, no attempt should be made to insert the hand to locate the afterbirth. If a small part is hanging out, it is best not to try to pull it, because the membranes are attached to the lining of the uterus by cotyledons, or buttons, and the delicate lining of the uterus may be torn.

Some ewes eat the afterbirth, which will do them no harm, and some move around so much while licking off their new lambs that the afterbirth becomes buried in the straw. So, the fact that you do not find an afterbirth is not necessarily very serious. Give the ewe all the warm water she will drink and some tasty hay; if she eats and drinks and allows the lamb to nurse, all is well. If not, and she seems feverish, veterinary help should be sought as soon as possible.

ANEMIA

This is one of the more commonly neglected diseases of lambs, caused by parasites or by deficiencies in the rations. The gums and eyelids will be pale, even almost chalky white, and the lambs will lose condition. Anemic lambs should be wormed with Thibenzole and then treated with injectable iron; supplement their rations with a stock tonic (available from a feed mill).

BLOAT

In sheep, bloat is especially dangerous because of the possibility of paralysis of the first stomach, or rumen. The problem normally occurs when sheep are grazing lush pastures, but may also be caused by some forms of poisoning. There is a rapid accumulation of gas in the paunch and reticulum, and the animal is unable to eliminate this by belching.

Some kinds of alfalfa cause a froth to form in the stomach, which blocks the gullet. The animal's left side will bulge abnormally, and respiration will be rapid and shallow. The condition is treated by the use of antiferments; Norton Laboratories has a liquid form of Bloat Guard that can be administered with a dose syringe. In emergencies, a dose of 2 tablespoons of kerosene in ½ cup of milk can be given from a narrow-necked bottle; a teaspoon of formalin in ½ cup of water may be used. Sometimes putting a stick across the mouth, behind the teeth, or just tickling the sheep under the chin will start them belching. In lambs, a few teaspoonfuls of liquid Pepto-Bismol will give good results.

If bloating is a problem, keep Bloat Guard on hand. To prevent this condition, give animals some hay before turning them onto rich pasture or onto pasture that is wet from rain or dew. Have water readily available, so that sheep drink a little at a time throughout the day, and not a great quantity all at once.

In severe cases a veterinarian may use a trocar to make a hole in the paunch to release the accumulated gas. This is not a do-it-yourself remedy, as there is danger of leakage from the rumen to the intraperitoneal spaces, leading to peritonitis. This tool should be used only by a veterinarian, as the very last resort.

CASEOUS LYMPHADENITIS
(LUNGERS, LUMPY JAW)

This is an infectious disease of sheep. It is chronic and is the reason many carcases are condemned in inspected slaughterhouses. Affected lymph nodes become large abscesses filled with greenish pus, usually under the jaw and on the neck and shoulder. Unlike the white pus of infected wounds, there is no odor. Infection enters through docking and shearing cuts and through wounds. The lungs, liver, and kidneys may become involved, and abscesses in the lungs cause chronic coughing and nasal discharge.

Sanitation is important in controlling the spread of the condition, and affected sheep should be isolated and the abscesses treated with iodine. Organic iodine, added to the salt, may be useful in flock treatment. When a shearer cuts into an abscess during shearing, the combs and cutters should be changed before going on to another sheep.

COCCIDIOSIS

Caused by a protozoa, this may occur in sheep after they are a few weeks old. It usually attacks lambs, and sometimes sheep, that are stabled in damp barns or sheds and whose fleeces pick up wet manure. Symptoms are copious and frequent bloody diarrhea, loss of weight and appetite, and listlessness.

The best cure is removal to a clean stable with dry bedding. Shearing the ewes will help. Sulfamerazine or sulfamethazine, three times a day for seven days, often corrects the situation. The addition of 1 pound of either of these drugs to 45 pounds of feed will also help. Good hygiene is the best preventive measure.

COLIC (ACUTE INDIGESTION)

This may come from eating frozen or wilted forage. The symptoms are much like bloat: the left flank may be distended, with the sheep stretched out as if trying to urinate. A teaspoonful each of powdered ginger and turpentine, mixed in a teaspoonful of honey and smeared over the lower teeth, will give relief in mild cases. Note, however, that turpentine must not be given to ewes heavy with lamb. For these, you can mix 20 drops of essence of peppermint in 2 tablespoons of milk of magnesia, with several tablespoonfuls of honey; smear this inside the lower teeth with a tongue depresser. If impaction (digestive blockage) is suspected, ½ pint of mineral oil may be given with the above.

CONJUNCTIVITIS (PINK EYE)

This is common among sheep and is self-limiting. The disease is spread by direct contact or by flies and may be controlled by prompt isolation of infected animals. The eye will be sore and runny and it may be closed. It will usually be inflamed, causing the sheep to be restless, reducing weight gains in lambs.

Treatment with ophthalmic ointment or Argyrol 15 percent solution will be helpful, but one of the best cures is the sheep's own tears. The eye should be inspected to make sure the irritation is not caused by a foreign body. (There is a hole under each eye, a gland, that is a natural part of the sheep's face.)

COUGH

A cough is a symptom of some disease or irritation, possibly worms, pneumonia, or dust and weed seeds in the throat. Sheep that are chronic coughers and snufflers should be culled, because they spread germs among the lambs.

DIARRHEA (SCOURS)

Like a cough, this too is a symptom of a disease. It may be caused by worms, diet, moldy feed, or poisoning. Animals afflicted with scours become debilitated by dehydration and lack of nourishment. Careful feeding, watering, and handling before and during shipment will help prevent shipping fever, another form of hemorrhagic septicemia, or scours; and vaccination two weeks before shipping will give added protection.

When diarrhea is caused by worms, the animals must be wormed and then treated for anemia. When the cause is indigestion, ½ cup of milk of magnesia may be given to clean out the system, followed by ¼ cup of Kaopectate. All animals suffering from diarrhea will benefit from a course of a good stock tonic to restore the natural balance of fluids, electrolytes, and minerals in the body. But, as the labels on bottles of over-the-counter medicine always advise, "If the condition persists more than two or three days, consult your physician."

EVERSION OF THE WOMB

This condition sometimes follows prolonged labor, or may follow the onset of milk fever, which causes the ligaments holding the womb in place to become weakened. The ewe lies on her side, weak and with cold ears, after lambing; the womb rolls out, looking like a reddish balloon. Expert assistance should be summoned, but if this is not available, the following emergency measures must be carried out.

Lay a clean cloth under the everted organ and pick off any dirt or bits of straw. Do not wash unless it is extremely dirty, as this seems to increase the straining. The ewe's hindquarters should be at least six inches higher than the head; this may be effected by putting a sheet on an old door and sliding the sheep onto it, and then raising the end of the door which is under the hindquarters of the sheep. With a very clean hand and arm, make a fist and push the organ back so that the outside will be on the inside. Once it is in, place a bale of straw under the ewe to support her middle and keep the hind parts elevated for several hours.

If the womb stays in for several hours, the chances are good that it will stay there. Harnesses for preventing recurrence of eversion are available from Sheepmen Supply Company, Barboursville, Virginia 22923, but as this trouble is likely to recur and is likely also to be hereditary, the affected sheep should be culled.

FOOT ROT (FUSIFORMIS NODOSUS) (INFECTIOUS PODODERMATITIS)

Of all the afflictions that shepherds have to contend with, one of the most aggravating is infectious pododermatitis, or foot rot. This is a highly contagious disease, usually found in wet, marshy areas and irrigated pastures. The first sign is lameness in one or more sheep. The foot will be found to be slightly swollen and the skin in the cleft between the toes will be puffy and moist. If not checked, the disease will spread under the hoof, causing the hoof to separate from the foot and resulting in a deformed foot. Pus oozes from the toes. The foot and lower leg will feel hot to the touch and there will be a foul smell. While foot rot is seldom fatal, it causes severe loss of condition; ewes produce less wool and milk, lambs become stunted, and rams may be useless at breeding time. The infection may linger in the sheeps' feet for years, but will die out of the soil in a few weeks if no sheep are put on the area. Sheep whose feet have been treated must be kept separate from those whose feet were not infected. As a precaution against infection, the uninfected sheep may be run through a foot bath of copper sulfate solution daily. This bath will also serve as a precaution against foot rot in marshy areas. Sheep don't like to go through water and will be more likely to go willingly through the foot bath if some straw is laid in the solution.

Treatment calls for careful and thorough trimming of the hooves, removing any horny growth curving under the foot or "sled runners" in front of the hoof (see the following section on *trimming*). This is a backaching and time-consuming job, but very important. Beware of the assistant who can trim large numbers of feet in a short space of time, because a badly done trimming job is useless. Affected sheep can be put in a foot bath containing a disinfectant made by mixing 2½ pounds of copper sulfate in a gallon of water, or one gallon of 38 percent formaldehyde with 9 gallons of water. Keep the sheep in the bath for at least twenty minutes. If the copper sulfate solution is used, the foot bath must be made of wood because the solution will be corrosive to metal. The foot bath treatment should be repeated weekly until all evidence of infection disappears.

When a small number of sheep are involved, good results can be obtained by trimming the hoof to remove diseased tissue, followed by an intramuscular injection of 5 cc. Combiotic on the first day and then 3 cc. twice a day until the infection is cured. Mild cases will respond to treatment with foot rot spray or Kopertox. Soak a rag in the solution and apply it to the trimmed hoof, then put the foot in an old sock and wrap the whole thing in a plastic bag that is long enough to reach up the leg. Bandage the leg to keep the bag on. By the time the sheep has managed to get all this off, the foot should be cured.

Not all sore feet are caused by foot rot. If a sheep is limping, inspect the foot to see whether the hooves are painfully overgrown. The hoof should not curl under the toes, nor should it curl up in front. Another frequent cause of lameness is a small stone or seed wedged in between the toes, or a split hoof resulting from an injury. There is a small gland between the toes of the hoof which produces a secretion that is important to the health of the foot. This is a normal part of the foot and should not be removed. There is no truth to the assertion that removing this gland will cure runny noses in sheep!

• *Trimming hooves.* Use foot-trimming shears or, in tough cases, farrier's hoof-nippers and a *sharp* pocketknife. If the horny part of the hoof is allowed to grow over the fleshy part, the sheep will begin first to walk on its heels and later to walk as little as possible. This leads to serious problems, because a sheep that cannot walk properly cannot eat properly and becomes undernourished and susceptible to worms or pneumonia. Ideally, there will be only a little trimming on the bottom and perhaps some on the tip.

Be careful when working on the tip of the hoof, as a vein and artery extend almost to the end of the living part of the hoof. If these are cut the sheep may bleed badly, and the cut should be treated with styptic powder. Pare carefully; when the hoof once more looks in normal shape, a band of translucent material will appear as you get close to the quick.

When the toes have been allowed to get very long the soft part becomes deformed, and inside the skin mantel the flesh will begin to decompose. This chalky white material can be scraped away quite easily and should be removed until you see either black flesh in the case of Suffolks or Hampshires or pink flesh in the case of Dorsets. This is healthy tissue, and if no infection is found, the job is done. If the foot is found to

Badly overgrown hoof.

Only moderate trimming needed.

be infected, the foot rot treatment should be given before the sheep is put back with the other animals.

Barnyard surgery must be undertaken with care. If blood is drawn while the toes are being pared or cleaned of infected matter, the hoof must be treated with antiseptic and protected with a covering to keep germs and further infection out of the wound. Because it is impossible to prevent the sheeps' feet from coming into contact with manure or with other germs in the ground during the course of normal grazing, steps must be taken to keep the foot covered until the flesh is completely healed, or further infection and deformity may result.

113

GRASS TETANY (HYPOMAGNESEMIA)

This ailment can occur any time there is a sudden change to rich green feed, and usually happens in spring when the sheep are first turned out on lush pasture. The sheep becomes weak, blind, and staggers aimlessly around until it falls down. Attacks can also be brought on by excitement or stress. When sheep are to be grazed on early growth of cereal grains or on very lush pasture, they should have access to a mineral mix containing 16 percent magnesium oxide. Feeding a little oats before turning out into pasture will help prevent this disease. Supply plenty of salt and put a handful of slaked lime in the water if a drinking trough is used.

Stricken animals must be treated immediately with an injection of 50 to 100 ml. of calcium borogluconate; veterinary advice should be sought in calculating the dose. Caldex, that handy pickup which is so good for weak lambs, contains magnesium chloride and calcium borogluconate combined with dextrose and is an effective remedy for grass staggers.

HEMORRHAGIC SEPTICEMIA
(SHIPPING FEVER, PASTEURELLOSIS)

See also *diarrhea*. This is one of the most important of the diseases that infect sheep. It usually attacks lambs follow-

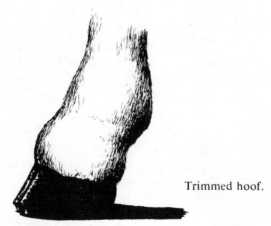

Trimmed hoof.

ing shipping or other stress and is characterized by fever, depression, and pneumonia. It is frequently fatal. Every effort should be made to avoid overcrowding when shipping, as well as hard driving and lack of rest, exposure to bad weather, or fright. The addition of sulfas in the feed at the rate of 100 mg. per head during stress periods may reduce the incidence of shipping fever. When disease strikes very young lambs, good results will be obtained with a dose of 2 cc. of penicillin or Combiotic, with Kaopectate to reduce diarrhea, and careful nursing.

IMPACTION

When a sheep stands around looking miserable, not eating, not chewing the cud, and with bad smelling breath, impaction (or constipation,) should be suspected. Any one of the stomachs may be infected, though the rumen is the most likely one, and impaction of the fourth stomach is the most serious. Use ½ cup of mineral oil to correct the situation; if it does not work, veterinary help is needed. Epsom salts and harsh physics may do more harm than good and can even cause miscarriage if a treated ewe is pregnant. Adequate exercise, nourishing food, salt, and water with the chill taken off will help to prevent this condition.

KETOSIS (PREGNANCY DISEASE)

This is a disease that strikes ewes in the later weeks of pregnancy and, unless it is detected early, is usually fatal. The cause is a disturbance of carbohydrate or sugar metabolism, resulting both in ketosis and a fatty liver: the livers of sheep suffering from pregnancy disease are crumbly and mottled with fat. The disease may be triggered by sudden stress of moving, storms, sudden changes of diet, or inadequate nutrition. The same condition may also occur in rams.

The early symptoms of pregnancy toxemia are seldom noticed. Often the first sign of trouble is the sight of one or more ewes standing apart from the rest of the flock, heads held high and not feeding. Such ewes are completely unconscious of their surroundings and will take no notice of anyone who approaches. There may be grinding of the teeth, discharge from the nose and mouth, and constipation. They may stand around for two or three days but will eventually go down, either resting on the chest with the head turned round near the flank, or on the side.

If the condition is noticed in time, early drenching with propylene glycol (2 ounces four times daily) will give good results. Treatment should be continued for some days after the ewes appear to have recovered. Should a ewe give birth to the lamb, recovery will be immediate.

The prevention of pregnancy disease is largely a matter of management. Ewes should be maintained in good flesh, neither over nor underweight; the ration must provide sufficient carbohydrates to keep them in good thrifty condition without making them fat. Molasses or brown sugar can be added to the ration, at the rate of 3 pounds per 100 pounds of feed. Sudden changes of diet, periods of starvation or stress, and overfeeding must be avoided.

LIP AND LEG DISEASE
(ULCERATIVE DERMATOSIS)

This is a contagious disease of sheep caused by a virus. It is spread by contact, but requires a break in the skin to enter.

There are raw sores on the skin of the lips, legs, feet, and genital organs. Rams may show a small ulcerative lesion of the prepuce, a condition very similar to sheath rot. The lesions caused by lip and leg disease are easily confused with those caused by soremouth, but when the scab is removed the lesions caused by soremouth protrude, while those caused by lip and leg disease are craterlike.

Infected animals should be isolated and kept out of the flock until the disease has run its course—anywhere from two to eight weeks. Treat as for foot rot or soremouth. If the foot rot treatment is used, the copper sulfate solution can be splashed on, or a paste may be made and dabbed on the sores.

MASTITIS (BLUE BAG)

When a ewe is restless, refuses to allow her lamb to nurse, and seems to be in pain, she should be caught and the udder examined. If this is hot and congested, or cold and clammy, treat the ewe at once. Usually only one side is affected. There may be a reddish secretion from the nipple because of hemmorrhaging inside, and the ewe will have a fever of 104° to 106°F.

There are many causes: the lamb may injure the udder by rough "bunting" when nursing; there may be cuts or other injuries; or the udder may be congested with too much milk if the ewe has just lost a lamb.

Affected ewes should be isolated and bedded on clean straw or clean ground. In some cases a strong, 15 gm. dose of sulfamethazine, given at the first sign of infection, and a second dose given twenty-four hours later, will bring about a cure, but treatment must be given promptly. If the condition is neglected an abscess may form and burst, with parts of the udder sloughing off. Once an abscess does form, antibiotics will usually prevent the spread of infection; should the abscess burst, the area can be treated with either tincture of iodine in glycerine, or any barbwire balsam.

Ewes may harbor the organism in their udders and lambs may spread the disease from one sheep to another, or ewes may pick up the infection from the ground. Sanitation and good management play a large part in preventing outbreaks of mastitis.

In mild cases, bathing the udder with a solution of 2 heaping tablespoonfuls of salt and the same of vinegar in a pan of warm water, or a solution of ½ cup of epsom salts in a pan of warm water, will help reduce discomfort and inflammation. A mixture of 1 ounce of pokeroot (a common weed) in 7 ounces of soap liniment should then be applied. Any other liniment that does not contain camphor may be used. (If the udder is to be dried up, camphorated oil may be applied twice daily.)

Partial milking and discontinuance of grain feeding will help dry up the udder. Any milk drawn from an affected ewe must be disposed of and is not fit to feed to the lamb. Always milk into a pan, not onto the bedding or the ground; if some milk does drip onto the bedding, it should be destroyed and fresh bedding put down, because the milk carries the mastitis organisms. Ewes with a tendency to mastitis or with faulty udders should not be carried over the next year's breeding season, even though the udder may appear to heal. Chances are that one or both sides will be blind and may abscess or be otherwise nonfunctional.

MILK FEVER
(HYPOCALCEMIA, LAMBING SICKNESS)

This is one of those exasperating conditions that may be confused with a variety of complaints, including mastitis, enterotoxemia, pregnancy disease, and hypomagnesemia. The disease usually occurs during the last weeks of pregnancy or just after lambing. The ewe takes no notice of her young, her ears are cold, and the entire body seems clammy. There is no fever, but bloating is common. Affected sheep appear to be in a coma.

The Caldex treatment is of benefit (see *grass tetany*). Avoidance of stress and upsetting conditions or sudden changes of feed will help prevent outbreaks.

RAM EPIDIDYMITIS AND
SHEATH ROT (PIZZLE ROT)

Both these conditions of the ram's genital organs are contagious and are likely to reduce the animal's fertility. One should never buy or use any ram that shows signs of swelling or lesions on the testicles, the sheath, or the prepuce.

SOREMOUTH (CONTAGIOUS ECTHYMA)

Soremouth is caused by a virus. Initially a reddened, pimplelike area develops, which later becomes bluish and fills with fluid, like a blister. Within hours the blister will break and a crusty scab will form. This usually occurs on the lips, and more rarely, on the eyelids, feet, and teats of ewes. The disease runs a course of three to four weeks and will eventually clear up, but if not checked it can spread through the flock to affect all the sheep, with the lambs most severely affected. The sores make lambs reluctant to nurse, and the teats of the ewe are too sore for her to permit the lamb to nurse, causing the lambs to lose weight and the ewe to develop an inflamed udder and possible mastitis because of congestion.

If a flock is known to have been exposed to infection, or if the condition exists in the whole flock, the animals should be vaccinated. The directions on the package call for vaccination on the inner surface of the rear leg, but better results have been achieved by pulling a tuft of wool off the breastbone and brushing the vaccine onto the open wool follicles. Grown sheep may be vaccinated on the under surface of the tail. Directions as to dosage must be followed carefully and, because this is a live virus, rubber gloves must be worn by the handler, with the greatest care taken not to get the vaccine on the handler's own skin. Once vaccination is started the program must be followed every year because a live virus has been introduced into the flock—so be sure you really need it. Note that soremouth is transmissible to humans (orf) and produces painful sores on hands and arms.

A Technician-Shepherd of the Ministry of Agriculture and Colonization in Quebec gives this alternate treatment for soremouth (reproduced by courtesy of *The Shepherd* magazine): "Make a good 'salad sauce' by mixing together vinegar, salt and pepper. Dip a piece of rag in the mixture and remove all the scabs, whether bleeding or not. Wash the lip wounds with the rag. From the start of this first treatment, the disease will be under control, but as a precaution give a second treatment two days later and a third two days after that.

"If the inside of the mouth is affected, add copper sulfate (bluestone) to the mixture (about half-a-teaspoon to a cup of vinegar) and wash the mouth wounds with the solution every day until recovery. To each cup of vinegar add a level teaspoon of salt and a full teaspoon of pepper, mixing well. Sometimes a nursing ewe's udder and teats will be attacked, and here the disease does not respond to the same treatment as the mouth. For this you will need about 2 or 3 pounds of poplar buds; crush them in a bowl with a pound of pure lard. Use a wooden pestle to crush a small quantity of buds at a time, adding the lard gradually. Always put the lard grease in the bowl first.

"When all the buds are crushed and mixed with the grease, put the mixture on the stove to cook gently for about two-and-a-half hours. Do not allow it to boil too much. When the unguent is cooked, allow it to cool, crush a square of camphor, and mix together. Strain the mixture through an old towel, twisting the towel to extract all the grease. When the grease has cooled, the ointment is ready. Use only pure lard because this is penetrating while vegetable and mineral grease is not. Rub the wounds of the affected udders until recovery, treating twice daily. Black poplar is the best kind; its buds are very sticky and are available from about November until May."

This ointment may be made up and stored in a cool place.

TETANUS (LOCKJAW)

The tetanus organism is widely distributed in cultivated soils and in horse dung. It is an anerobe (able to live without oxygen) and usually causes trouble when it gets into a wound which then closes up and heals. In the absence of oxygen, the spore grows and liberates a powerful toxin that causes the animal to suffer violent spasms, labored respiration, and rigidity.

Once the disease reaches the stage in which the animal cannot move or swallow, the chances for recovery are not very good, but some cases will respond to careful nursing. The sheep or lamb should be isolated and bedded in a clean and comfortable place. Every few hours a little nourishment can be dribbled down the throat, being careful not to let any get into the lungs. For a lamb, milk will do, given every two hours, with an occasional sip of water. An adult sheep can be fed honey and water or glucose and water in the same way. At the time of feeding, stand the animal on its feet and rub it a little, then lay it down on its other side, to avoid having all the contents of the stomach "settle" on one side, and to keep the circulation going.

If the sheep is going to recover, it will show signs of doing so within a few days; at first it will nibble a little at whatever it can reach, and then will gradually reach further and further as the stiffness relaxes. This interest in nibbling helps recovery, and it is a good idea—if the animal is small enough to be carried—to take it outside and put in on clean grass for a short time. As the ability to feed itself improves, the milk or honey-water feedings can be more widely spaced.

Prevention of tetanus infection should be a consideration whenever an animal has an open wound, and at the time of docking and castrating. Keep the area clean, and give 0.5 cc. of tetanus antitoxin. Lambs that are to be docked or castrated by the rubber ring method should also receive the antitoxin.

URINARY CALCULI

Urinary calculi are stones in the bladder. The rams are usually the only animals affected, as ewes seem to be able to pass the stones. No one knows exactly why rams get stones, but it is thought that they may be the result of high potassium intake, a high proportion of beet pulp or grain sorghum in the ration, or a high level of phosphorus in combination with a low level of calcium in the ration. The sheep will make frequent attempts to urinate and will be in pain. The bladder may rupture, or uremic poisoning may set in. Once the calculi have developed, dietary treatment does not do much good; muscle relaxants may allow the passage of the stones if used before the bladder ruptures. The condition is unlikely to occur if 1 to 3 percent salt is added to the grain ration, using the higher rate in winter when water intake is usually lower.

WOOL BALLS

When a lamb stands around looking miserable and does not eat, for no apparent reason, treatment to remove wool balls might bring about a recovery once other causes have been considered and rejected. Some lambs eat wool, either from a dietary deficiency or because they got the taste for it when searching under their unshorn mothers for something to eat, and this wool forms balls in the stomach. Administering ½ cup of mineral oil will help the lamb expel the obstruction.

INTERNAL PARASITES

ROUND, OR BLOOD-SUCKING WORMS

There are several genera:

Large stomach worm, barber pole worm *(Haemonchus)*
Brown stomach worm *(Ostertagia)*
Stomach hairworm *(Trichostrongylus)*
Thread-necked worm *(Nematodirus)*
Hookworm *(Buonostomum)*
Nodular worm *(Oesophogostomum)*
Large-mouth bowel worm *(Chabertia)*
Whipworm *(Trichuris)*

The most important symptom is anemia because, for all their impressive-sounding names, these worms are in effect bloodsuckers and can cause severe anemia which produces weakness, low milk production, dull wool, and poor condition.

Inspection of the mucous membranes of the mouth and the lining inside the eyelids will show these to be pale pink or chalky white, depending on the severity of the infestation. Bottlejaw, a mumpslike swelling of the lower jaw, is sometimes seen in severely parasitized sheep, and there may also be diarrhea.

Adult roundworms live in some part of the alimentary tract where, following mating, the female deposits eggs that are passed in the droppings. These eggs develop into larvae that hatch within a few hours. The larvae use moisture to move away from the manure, and reach an infective state within two weeks. In warm, humid weather the spread of possible worm infestation can be greatly reduced by not permitting the sheep on the pasture until the sun has dried the dew. When sheep eat the larvae off contaminated grass or feed, these larvae mature in the digestive tract and produce eggs that hatch, and then the worms attach themselves to capillaries of the gut wall and suck blood from them.

The worms do not compete with the sheep for food. The harm they do is caused by loss of blood and consequent anemia of the host animal. Nodular worms penetrate the gut lining and surround themselves with masses of fibrous tissue, which prevent absorption of digested food by the sheep.

Treatment is directed at destroying the adult worms, destroying the eggs and larvae, and breaking the reproductive cycle. Some species of worms are resistant to the effect of certain drugs, so it is important to take a fecal sample to the vet so that you know which species you are dealing with.

Because the worms are bloodsuckers, it is necessary to treat the anemia as well as the worm infestation. Severely parasitized sheep have been treated with transfusions of whole blood, but as this procedure is not practical for the average flock owner, doses of injectable vitamin B can be used.

• *Worm medicines used in the control of roundworms.* Phenothiazine is the principal vermifuge used for sheep. The full dose of this drug should be given at one time, as efficiency is lowered by splitting the dose over several days and incorporating it into the feed. Colloidal suspensions of very finely ground powder are better than pills or boluses. Phenothiazine has an unpleasant taste, and the sheep will very likely refuse to eat food medicated with the drug. It is the most effective drug for use against the common stomach worm, which usually reaches its peak during hot humid weather. The standard dose is 25 gm. for mature sheep and 12.5 gm.

for lambs under sixty pounds. Check the label on the container to be sure you are using the right dose, and use a standard 4-ounce dose syringe for treating small flocks. An automatic syringe will simplify the job when worming larger flocks. Do not use the drug on ewes in late pregnancy. Be cautious about dosing white-skinned sheep in very hot weather, as this drug may produce a photosensitivity reaction (sunburn); such sheep should be kept indoors or in deep shade for twenty-four hours after worming. Never combine two worm medicines in the hope of improving efficiency—the dose may kill the sheep.

• *Phenothiazine-salt mixtures.* Pheno-salt is usually mixed in the ratio of 1 pound of phenothiazine to 9 pounds of salt, but it has been found that sheep will consume more salt when a ratio of 1 pound phenothiazine to 14 pounds salt is offered. This mixture is also helpful where lungworm is a problem; the drug seems to make the lungworm larvae inactive before they are passed from the sheep.

• *Thibendazole (Thibenzole).* This drug is effective for the removal of blood-sucking worms. It has many advantages: it is tasteless, has a wide range of efficiency and safety (overdosing will not harm the sheep), and is safe for ewes in late pregnancy and for use on animals in weakened condition. The drug is more expensive than others used for worming sheep, but this disadvantage is largely offset by the safety and convenience of the product.

Thibenzole is also obtainable in feed, sold as "cattle wormer" feed, and can be fed to the sheep in the quantities recommended on the container. If the animals clean up all the feed immediately, good results can be expected, but if they dislike the taste and refuse it, then dosing with the powder or a suspension of powder in water will become necessary. The drug is also sold in boluses. If the powder is mixed with water to make a liquid dose, shake the container every time the syringe is filled to keep the powder in even suspension.

• *Haloxon (Loxon).* This is an organic phosphate drug that is relatively safe and highly efficient against blood-sucking worms of sheep. It is economical to use. Correct dosage must be administered and the label directions followed carefully.

• *I-Tetramisole (Tramisol)*. This soluble drench powder is effective against blood-sucking worms and lungworms. It is easy to mix and easy to use, but is more toxic than other commonly used drugs, and dosage must be measured very carefully.

• *Paritrope*. This is a liquid wormer, ready to use without dilution. It is very effective against stomach worms, tapeworms, and coccidiosis. It may be used on young animals, and pregnant and debilitated sheep. It contains body-building minerals to aid in restoring depleted supplies caused by parasite infestation.

LUNGWORMS

Lungworms are a serious problem in some sheep flocks. The adult lungworm lives in the air passages of the lungs, causing severe irritation and deep coughing. The eggs are coughed up, and either eggs or larvae are passed in nasal discharges or in the droppings, to be eaten by the sheep if the feed is contaminated by the droppings. The larval worms then penetrate the digestive tract and migrate to the lungs, where they mature. Dark and damp conditions are ideal for propagating lungworms and infecting baby lambs. Tramisol is the most effective drug available for the treatment of lungworms.

TAPEWORM (MONEZIA EXPANSA)

Signs of tapeworm infestation can easily be seen in the droppings as white segments, and they are often the only indication of worm infestation that causes any alarm. The tapeworm is, in fact, much less harmful than the stomach worm, and studies have shown that if lambs and sheep are inoculated against enterotoxemia, tapeworm infestation will not cause serious loss of condition (see *enterotoxemia* in chapter 4.)

Well-nourished sheep in good condition do not suffer severely if infested with tapeworms, and the condition is self-limiting as long as the worm load does not become so heavy as to cause intestinal blockage. The eggs and segments pass out with the feces and are ingested by an intermediate host, the beetle mite or oribatid mite, where they develop into a larva known as a cysticercoid. The sheep swallow the larvae, which travel to the small intestine, where the adult worms develop. Control is made more difficult by the fact that the larvae are protected by the mites, which can withstand severe conditions of cold or heat.

The larvae climb up grass blades during the cool of early morning, and then drop back to the protective soil as the day becomes hotter. Keeping lambs off pasture until the sun has warmed the grass provides a measure of control.

Lead-arsenate mixed with phenothiazine will control tapeworms, but this is a dangerous medicine and should not be used more than twice a year. Less risky is Ovine Cu-Sate or Paritrope, but these must also be used with caution. Ovine Cu-Sate boluses should be dipped in oil to make them slide down more easily; use one bolus per adult sheep, half for lambs under sixty pounds. Arsenic builds up in the liver and should be used with great care, and very sparingly. The best defense against tapeworm is good nutrition, protection against enterotoxemia, and keeping the lambs and sheep off damp grass.

LIVER FLUKE

The liver fluke requires an intermediate host for reproduction and is carried by certain kinds of snails. Keeping the sheep out of wet spots and marshy areas where snails may live is an effective means of control. Treatment for liver fluke must be administered by a veterinarian.

For their size, parasites are the hungriest and most expensive animals to feed on the farm. They rate as the biggest disease problem of sheep. As flock size increases and the use of grazing land intensifies, so the parasites multiply, causing serious losses, especially in farm flocks. Infestation may cause the death of the animal, but in most cases it causes anemia, loss of condition, poor growth, and ragged wool. To deal effectively with the problem it is necessary to know which parasites you are dealing with and what methods will control them.

Worm build-up will be greater in warm, humid weather and on damp ground. This may explain why it is that sheep sometimes appear to do better during a hot dry spell, when the pasture is not very good, than they do during damp weather when the pasture grows quickly and is lush. The worm eggs are able to last much longer in a damp environment than when exposed to hot sunlight.

The symptoms of many different kinds of worm infestation are similar. The best way to identify the culprits is to take several samples of droppings to the veterinarian's office for analysis. Waiting until an animal is dead and then ordering a post mortem does little good because decomposition of dead animals is very rapid in warm weather. Fecal samples should be put in an airtight container to avoid loss of moisture and taken for analysis immediately.

Normal control measures call for drenching of ewes in spring before they are put on pasture, at midsummer as they are rotated to clean pastures, and again in fall during the flushing period. By using a different product for each worming, you prevent the build-up of resistance to the drug used.

Good management is essential for effective control of worms. It is useless to worm sheep indiscriminately with drugs and then put them right back under the same conditions as existed when the worm build-up occurred. Nutrition, too, plays a large part in enabling sheep to resist worm infestation; a well-nourished, healthy sheep will suffer to a far less degree from worms than will a thin, weak animal. If the flock looks unthrifty and sad, attention should be given to their living conditions before reaching for the worm pills—perhaps the sheep are suffering from low-level starvation or lack of clean drinking water, salt, or minerals.

Attention to the following precautions will help prevent parasitic infestation:

1. Provide clean, rested pasture for lambs. Lambs are more susceptible to worms than are older sheep.
2. Keep all feed and hay off the ground; keep grain

troughs off the ground so that the lambs don't play in them as if they were sandboxes.

3. Feed the sheep well, as poorly nourished animals are an easy target for worms.

4. Install water tanks so that they do not leak and so that the surrounding area is well drained. Worm larvae love moisture.

5. Keep pastures clipped, where possible: sunlight kills worm larvae.

6. Add a low-level wormer to the salt ration and have it available to the sheep at all times. Keep salt boxes covered, because salt picks up moisture from the air and will not be appetizing if it is damp.

Two low-level wormers that have been shown to be effective are phenothiazine and diatomaceous earth. Phenothiazine, when finely ground and mixed with salt, provides a reliable and effective worm control. It inhibits the development of larvae in the feces and consequently reduces the build-up in heavily infested areas. There is some staining of the wool around the dock when phenothiazine is used, and evidence suggests that worms may build up a resistance. Use 1 pound of phenothiazine mixed with 14 pounds of salt.

The second method is the mixing of diatomaceous earth with the salt ration. Again, see the list of worm medicines for dosage. Low-level worming is not a substitute for regular drenching where sheep are infested; it is used in conjunction with drenching.

Diatomaceous earth is the fossilized remains of countless trillions of minute one-celled ocean plants called diatoms. The diatoms constructed tiny shells out of the silica they extracted from the waters. When a diatom died, its microscopic shell was deposited on the floor of the ancient seas in layers which became, in places, thousands of feet thick.

As the waters receded these shells were covered and became fossilized and compressed into a chalklike rock called diatomaceous earth. When this earth is milled, ground, passed through a screen, and put through a centrifuge, it becomes a fine talclike powder that can be handled safely with bare hands, fed to animals with no harm, and yet will kill insects on contact.

The razor-sharp pieces of diatom shell cut the skeletons of insects and parasites that, unlike birds and mammals, have their skeletons on the outside. When this outer covering is broken, allowing the body fluids to escape, the insect dies by dehydration. An internal parasite encountering diatoms in the sheep's gut is killed the same way, mechanically and non-chemically.

A box of diatomaceous earth can be placed alongside the salt box, or better, the salt and earth can be mixed fifty-fifty by weight. If mixed with the feed, 1 to 2 percent of the total animal ration is used.

Diatomaceous earth is nontoxic and contains manganese, magnesium, iron, titanium, calcium, and silicon. In addition to killing the worms inside the sheep, the diatoms excreted in the manure seem to keep down the number of flies and to reduce odors. Sheep like the salt mix and eat it freely. When the fossilized flour is mixed with a grain feed there is a certain amount of dust, which can be controlled by mixing the flour with a small quantity of molasses-coated grains, such as are used in horse feed, and then mixing this with the regular grain ration. The material can also be sprinkled over barn floors and sills to help keep down the fly population.

Sold under the trade name Perma-Guard, the product is available from Life-Guard Products, 1701 East Elwood Street, Phoenix, Arizona 85040.

DRENCHING AND WORMING WITH PILLS OR BOLUSES

Whatever method is used or product employed, worming and drenching must be carried out quietly and carefully, with strict attention to correct dosage. The sheep should be penned closely so that there is not a lot of chasing and catching, and young lambs and infirm sheep should be penned separately to avoid injury. Sheep should be marked with a crayon or paint dauber as soon as the medicine is swallowed so that none gets a double dose. If marking is done, the ones that have been wormed can be separated if all the sheep break away before the job is finished.

To give a liquid drench, stand just behind the shoulder of the sheep, place one hand under the jaw and with the other hand insert the nozzle of the drench gun (available from livestock supply houses) into the side of the mouth and over the tongue. Slide the nozzle back over the hump of the tongue until the end is just opposite the molar teeth and slanted slightly towards the side of the mouth—but don't force it down the throat. Always hold the sheep's head level, because the animal cannot swallow if the head is pulled back with the nose pointed up or to the side. Press the plunger gently and let the fluid trickle in.

It is important to see that the tip of the drench nozzle or balling gun is rounded and free of sharp edges. All equipment should be in top condition: completely clean and dry each piece after use, and dip the leathers on dose syringes in mineral oil before storing. Before starting to drench, check the syringes for accuracy of dose delivery. Proper worm medication depends on the use of the dose prescribed and its proper delivery to the right destination. A syringe pushed too far into the mouth may cause the liquid to run into the lungs, or, if the syringe is not inserted far enough, much of the medicine may be wasted.

A bolus will slide down more easily if it is first dipped in oil. Hold the sheep's head level, in the same way as if to drench, and slide the balling gun (available from livestock supply houses) in over the tongue. When it is about level with the molars, express the bolus and withdraw the gun. Hold the sheep's mouth shut for a moment—it's amazing how good some of them are at coughing up a bad-tasting pill.

Drenching should not be undertaken when the sheep are exhausted, nor in the heat of the day. At least two days should be allowed between drenching and treatment for external parasites.

WHY ARE SHEEP SUSCEPTIBLE TO WORMS?

Like death and taxes, worms are always with us. They come with the sheep, and if not controlled will multiply more or less rapidly depending on soil conditions, weather, and rate of stocking of the pasture.

When sheep are introduced to pasture where there have been no sheep before them, they thrive. This encourages the flock owner to increase the flock, and here the trouble starts. In the first place, the sheep is a reservoir of worms, even if regularly wormed, and it is a rare sheep that does not carry some sort of worm in it with which to contaminate a new pasture. When only a few sheep are carried on a large pasture, the worm larvae die out in a matter of weeks because the sheep have gone on to new and fresher grazing, but as the stocking rate is increased, sheep will be grazing the same area more frequently and are more likely to ingest the worm larvae that are lying on the ground. A heavily infested ewe may carry in her several thousand stomach worms *(Haemonchus contortus)*, and each female worm lays more than five thousand eggs every two hours, so that even a lightly infested sheep can start an infestation in a pasture.

Eradication of worm infestation depends on breaking their life cycle by sanitation, pasture management, and proper medication. When the eggs are passed in the droppings, they develop and hatch on the pasture or on the bedding in the barn or barnyard. The larvae use moisture to make their way away from the droppings; for about a week they will feed and grow, and then, having reached the infective stage, the larvae, enclosed in a protective sheath, will climb up blades of grass. The larvae then live on the food material stored in the body and, if not eaten by a sheep, will die in time. A larva in a sheltered place might live for months, but most will die off within a few weeks. When the larvae are eaten it takes about three weeks for them to develop into adult worms.

Well-nourished sheep in good condition can withstand worm infestation much better than the poorly nourished ones, and may even build up an immunity. Resistance is lower during times of stress, such as lambing and lactation, or when being shipped.

Regular seasonal drenching, spring and fall, will kill the worms in the sheep. It is necessary to know what kind of worms are involved before proceeding with the drench, as different medicines act on different kinds of worms. Immediately after the sheep are drenched they should be confined in the barn for twenty-four hours to avoid further contamination of the pastures with the expelled worms and larvae.

No drench is one-hundred-percent perfect. Some worms or larvae will survive, and in addition, some sheep seem to be able to cough up the medicine from hidden depths and spit it out after you are sure that it was swallowed. If the sheep has already been released, this rejected dose may not be noticed.

The value of feeding the phenothiazine-salt mix lies in the power of the drug to inhibit the development of the eggs into larvae on the pasture. The diatomaceous earth-salt mix acts to destroy the larvae and adult worms in the sheep's stomach.

Properly controlled pasture rotation is one of the most effective ways of limiting worm infestation. A worm egg requires two weeks to develop into a larva, and a larva requires about three weeks to develop into an adult worm. If the sheep are prevented from eating the larvae by moving them from the area after two weeks grazing, and the area is then rested for three weeks, the worm cycle will be broken. Short, dry grass provides a less favorable environment for the development of the larvae, and bright sunlight will lead to a much slower development, so pastures that are lush and long will benefit from clipping to keep the herbage short and to allow the sunlight to get to all the plants. Grazing cattle with sheep will increase the amount of pasture for both, as the cattle eat the longer grass and many of the tougher plants that the sheep ignore.

EXTERNAL PARASITES

Sheep infested with external parasites are unthrifty, slow to gain, and are restless and uncomfortable. They will rub themselves against gates and fences and may roll on their backs because of the itching. A sheep heavy in lamb cannot always right itself if on its back and may die if not helped.

The rubbing and scratching causes patches of wool to fall off and reduces the value of the fleece at shearing time. The skin of a sheep infested with ticks will develop cockles, small hard nodules that are the result of the ticks repeatedly puncturing the skin. These skins, which otherwise could be used for coats, gloves, and bags, must be used for low-cost shoe linings where the defects will not be noticed.

SHEEP KED

The most common external parasite is the sheep ked, commonly called sheep tick, which is a blood-sucking fly. Its entire life is spent on the sheep. One larva at a time is developed within the body of the female ked, and each female deposits a mature larva in the fleece about every ten days. These larvae immediately pupate, and in about twenty days adults emerge and begin to feed—and so the cycle continues. Each ked engorges with blood every twenty-four to thirty-six hours, and a heavily infested sheep will lose so much blood that anemia will develop. Wool growth is retarded and the gaining rate of lambs is slowed.

Fortunately sheep keds are relatively easy to eradicate. Application of an insecticide soon after shearing can be done with little effort and expense, and normally one application per year is sufficient. All the animals must be treated because the keds will leave the adult sheep after shearing and attach themselves to the young lambs. Sheep keds are not disease carriers and are not harmful to humans.

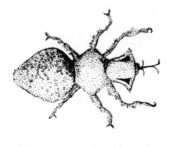

Sheep ked.

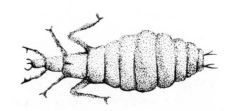

Lice.

LICE

Three species of lice occasionally affect sheep. Two are sucking and one is a biting louse; all cause intense irritation. Eggs are glued to the wool, and hatching occurs in five to eighteen days. The cycle of egg-laying and growth is completed in about two weeks. Control as for sheep keds.

METHODS OF TREATING SHEEP KEDS AND LICE

Until comparatively recently, the accepted method of ridding sheep of external parasites was by dipping into or running through a special tank. This involved either digging a pit and installing a six-foot-deep tank, with ramps and draining areas, or hiring a custom sheep dipper with his mobile unit. Newly developed insecticides have made a much easier job of this very necessary chore. When used as recommended, insecticides are efficient and safe, but some of them can be dangerous if improperly used. Read the directions on the container and do not improvise. If one ounce is good, two ounces are not necessarily better, and quite likely may be fatal. The following methods and preparations for treating external parasites are safe and simple.

• *Korlan 24E*. Before starting to work, pen the sheep in the barn or shed, or a corner of the yard, then set up small pens leading from the first pen. Alongside the second small pen, assemble the spray, water, some dry clean cloths (to wipe off any spray that the sheep may splash on you as they shake), and a measuring container. Don't use the two-cup Pyrex measure from the kitchen because this stuff is poison, and the container must be used for no other purpose. The whole job will go very easily if one helper allows eight sheep at a time into the first holding pen, while the second helper is responsible for marking each sheep with a crayon as he moves it into the second holding pen. The second helper can also keep the sprinkling can filled with the proper concentrate of insecticide.

The spray is made by mixing 8 fluid ounces of Korlan 24E in 3 gallons of water. Use 1 quart of solution per adult sheep. With the first eight sheep closely penned, pour the solution

down their backs, using a two-gallon garden sprinkling can. Immediately release them and run them out into
the fresh air. The liquid sinks into the bloodstream through
the skin, and keds sucking the sheep's blood are killed as they
feed. Developing pupae are also killed because the Korlan
has a residual effect lasting three weeks. Pregnant ewes may
be treated until within two weeks of delivery, but nursing
ewes should not be treated until the lambs are at least a
month old. Sheep must not be slaughtered for food within
thirty days of spraying. The barn or shed where the operation is carried out must be well ventilated, and workers
should be careful to get as little of the spray as possible on
their skin and to change clothes when the job is finished. The
sheep may appear to be intoxicated and stagger around after
treatment, but this will soon wear off.

•*Diazinon.* This treatment may also be applied by the above
method. It is an emulsifiable concentrate and is applied at
the rate of 1 quart of solution per animal, using ½ ounce of
50% Diazinon wettable powder per 3 gallons of water.

•*Coumaphos (Co-Ral).* This is used in solution of 4 pounds of
Co-Ral per 100 gallons water to wet down the animal. Co-Ral
may also be dissolved in mineral oil at the rate of 1 pound
Co-Ral dissolved in 1 gallon of mineral oil, and applied to the
sheeps' backs at the rate of 1 ounce per 100 pounds of body
weight. When this method is used, the oil may be brushed on
the sheep directly after shearing.

•*Marlate.* Marlate is a powder that can be applied directly
after shearing by taking up a handful and spreading it down
the sheep's back. Repeat in two weeks if necessary.

In her book *Herbal Handbook for Farm and Stable,* (Rodale
Press, Emmaus, Pennsylvania, 1976), Juliette Levy recommends either dusting with powdered derris root (derris is the
basis for rotenone dust) or making a dip with derris to which
a little oil of eucalyptus has been added. The solution should
be rubbed over the sheep immediately after shearing and applied to the lambs at the same time.

Do not spray or dust animals for ten days before or after
shipping or weaning or after exposure to disease. Do not
spray or dust animals within thirty days of slaughter for

food. Do not apply in conjunction with oral medication. Do not apply Korlan (Ronnel) or Co-Ral to animals receiving organic phosphate from other sources, and do not dip at the same time as worming.

WOOL MAGGOTS

Blow flies attack sheep having open wounds, fleeces bloody from lambing, or wool soiled with feces. The adult flies lay their eggs on the soiled wool and the maggots soon hatch and begin to feed. Open sores infested with maggots can become infected with bacteria, causing the death of the sheep.

Affected sheep should be sheared around the fly-struck areas and the wound and surrounding area treated with a special smear, such as Smear 62 (diphenylamine), EQ-335 (lindane and pine tar), or 5-percent Ronnel (Korlan). (For other measures, see *wounds,* the second section in this chapter.)

NOSE BOTS (HEAD BOTS, NOSE FLIES)

These are the maggots produced by a grayish colored fly, slightly larger than the housefly. The sheep nasal fly deposits larvae, not eggs, around the sheep's nostrils. The larvae crawl up into the frontal sinuses where they complete their development, and then migrate from the sinuses back through the nasal passages and drop to the ground to burrow into the soil and pupate. Anywhere from three to eight weeks later, the flies will emerge from the pupal cases, crawl

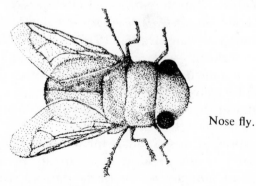

Nose fly.

to the surface, and find another sheep. While the flies are around the sheeps' noses to deposit their larvae, they cause great irritation; sheep rub their noses on the ground or against other sheep, and seek shelter in a cool, dark place. The most easily recognized symptom of infestation is a snotty nose. Affected sheep sneeze freqently and will cease to feed.

There is no absolutely safe and reliable treatment for nose bots, but the following three treatments will give good results.

• *Ruelene Drench.* Use a 21-percent emulsion, 2 cc. per ten pounds body weight, of Ruelene 25E. The treatment will make the sheep cough, and the animals will appear depressed for as long as twenty-four hours. Do not confuse this product with Korlan 24E, which is a highly toxic substance for external use only.

• *One percent Vapona (dichlorvos).* This ointment is custom mixed in petrolatum. Dosage is one to 1½ grams in each nostril of affected sheep. Repeat in two weeks. This medicine is 95 percent effective after two treatments.

• *Co-Ral Pour-On.* Wet down closely shorn sheep with a solution of 4 pounds of Co-Ral per 100 gallons water.

PSOROPTIC MANGE (SCAB OR SCABIES)

The common scab mite is one of the most destructive of the external parasites and also one of the most difficult to control. The mites feed upon the skin and suck out fluid. The wounds then become inflamed, and a crust forms. The wool falls off and the sheep will look ragged and miserable. Sheep scab is a reportable disease, and a qualified veterinarian should be notified immediately if you suspect that your sheep are affected.

7

Wool
and
Sheepskins

The production of wool in this country started when a group of combers and carders from Yorkshire, England, settled in the Massachusetts Bay Colony. These people became famous for their proficiency in making woolen cloth and were the first to build a fulling mill in this country. Fulling is a process of steaming, cleansing, and pressing the cloth to thicken it. Because there were no native sheep, the first wool was supplied by the English. Some sheep were imported from England, but they were unsuited to the harsh climate of the winters in New England and did not thrive. There was a shortage of wool until supplies became available from sheep that came from Spain into this country via Mexico.

At first, wool manufacture in the Colonies developed as a household craft with tools that the early settlers brought with them from England. The fleeces were washed in tubs and then hand carded and combed for the spinning wheel. Weaving was done on large, heavy looms that were built by local carpenters and operated in most cases by men working in shops or in sheds at the homes of their customers.

The grade of cloth produced by these weavers was coarse and uneven and in need of extensive finishing. Before the establishment of the fulling mills, the cloth was washed in warm soapsuds and then laid on a wooden platform and beaten with sticks. The fulling mill drew the pieces of woven cloth, which had been tied together end to end, between two rollers, which were set in a trough of soapsuds, and as the cloth was washed and pressed, the amount of shrinkage depended on the consistency of the soap solution as well as the amount of fullers earth added and the timing of the process.

At first the fleeces were dyed in tubs before spinning. Peddlers travelled round the countryside with dyestuffs, and housewives used plant dyes that they made themselves from plants familiar to them, such as oak, walnut, and acorn. Later, the fullers began to dye pieces after fulling, a process that required some knowledge and which gradually became performed by a separate group. Protruding fibers were snipped by hand from the finished fabric, and a nap was produced by hand carding. This nap was then evened off by more hand trimming with large shears.

The only native craft being carried on with wool at this time was the rug weaving done by Indians of the Pueblo tribes. These people learned their craft from the Spaniards who came to North America from Mexico. Most of the work was done by the women on primitive looms that could be easily dismantled and moved. The Navajo became well-known for their proficiency in livestock raising, and today their people still weave beautiful rugs from the wool of Navajo sheep. Native dyes are used, and the patterns are made up as the rug progresses, with no two designs being exactly the same.

Following the slow development of wool manufacture in the Colonies came a gradual transition from the household and handicraft system to that of the small factory. Most of the wool worked came from domestic sources and was usually very coarse, although some sheep had been imported from Spain for crossing with those already in this country. The early mills had a hard time, and it was not until 1794, when Messrs. Du Pont de Nemours and Delessert brought over a full-blooded Merino ram named Don Pedro, that some effort was made to improve the wool-producing qualities of sheep. Then, Robert Livingstone and David

Humphries brought in a flock of twenty-one Rambouillet rams and seventy-five ewes from Spain, causing a great flurry in wool speculation and prices, but still not much improvement in wool quality. In 1810, the American Consul in Portugal shipped over some four thousand sheep. Other importations followed, making a total of twenty-five thousand sheep within the next year. These were Merinos and laid the basis for a supply of fine wool. Interest in wool growing and manufacturing increased. Some English long-wool breeds were brought in, and traders and dealers in wool appeared all over the country.

As the wool industry grew, so new machines were developed, and imported machinery from England was modified to suit United States requirements. Types of wool available changed as the numbers of sheep increased. When sheep were bred for mutton production, Leicester rams were mated with Merino ewes, producing a good meat carcase but much coarser wool. Subsequent interbreeding yielded various grades of wool and led to the development of the dual-purpose sheep. There was also a demand for fine worsted—a fabric made by spinning long, fine-staple wool into a smooth thread—and woolen goods, leading to the development of the Delaine wools, which are the best type of Merino combing (long-staple) wools grown in the United States. Today's Rambouillet sheep are the descendants of these Merinos and are the sheep producing the best grade of fine wool in this country.

While the wool trade centered in Boston, and the mill industry became one of the major industries of New England, the carpet weaving industry grew up in Pennsylvania, and Philadelphia became the center of the carpet weaving markets.

After World War II the development of synthetic fibers dealt a heavy blow to the wool business. Everyone wanted wash-and-wear, no iron, no press, no fuss. Throughout New England and the Northeast, the once busy woolen mills became empty, vandalized structures. Machinery was dismantled, some of it to be exported to Russia and Japan, whose woolen goods are imported by this country in increasing quantities. Research on wool in the United States practically ceased, and the wool processing industry switched its interest to the processing of fibers currently in demand.

Just about the time that the man-made fiber industry seemed to have gained supremacy over the once-great wool business, Australia, a nation that for generations has been reputed to live off its sheep, together with her British Commonwealth partners engaged in research rescue operations for wool. The C.S.I.R.O. (Commonwealth Scientific Industrial Research Organization), through revolutionary research developments, showed that chemistry was not the exclusive domain of the synthetics industry. Wool fibers were modified by chemical developments to give them the qualities the public demands.

With this research to encourage them, the Commonwealth countries chipped in financially to form the I.W.S. (International Wool Secretariat), a body with laboratories ready to aid wool processors in making use of new developments. The "Woolmark," trademark of pure wool content, was devised by the I.W.S. to give pure wool an image that has since impressed fabric users all over the world.

It is due to the combined efforts of the C.S.I.R.O. and I.W.S. that wool research has been resumed in this country. When things were at their worst, the Australians came through, and it is because of their efforts that within the last few years there has been a realization that wool is the true miracle fiber, for both serviceability and beauty.

QUALITIES OF WOOL

To meet the requirements for top grade, wool must have a definite degree of fineness and must meet certain standards for length. Besides these, good wool has strength, elasticity, purity, character, uniformity, softness, and desirable color. A reasonable amount of grease and foreign matter is acceptable in fleeces and will not affect the price much, except in the case of very large flocks of sheep carrying animals whose wool is heavy in grease and thus shrinks more in scouring.

Fine wools shrink more than wools from half-blood fleeces, such as those carried by Dorsets and Corriedales, and range wools shrink more than farm-flock wools.

THE BLOOD SYSTEM

Wool is graded by the blood system of dividing wool into six grades. These are fine, half-blood, ⅜ blood, ¾ blood, low-quarter blood, and common (braid). Originally these fractional blood names denoted the amount of Merino blood in the sheep producing the wool, but now merely indicate the diameter of the wool fiber. This system is in turn giving way to numerical count, which divides all wool into fourteen grades, each designated by a number. These numbers range from eighties for the finest wool down to thirty-sixes for the coarsest.

The two general classes of apparel wool—as opposed to rug wool—are combing and clothing. By and large, combing

Wedding ring shawl, Zetland County
Museum, Lerwick, Shetland Islands.

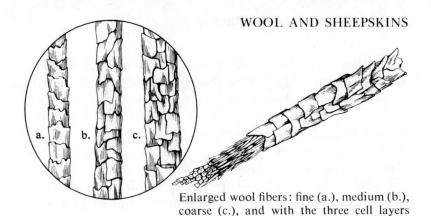

Enlarged wool fibers: fine (a.), medium (b.), coarse (c.), and with the three cell layers revealed.

wools are of longer staple than clothing wools, and clothing wool would be used for the production of woolen goods, while the combing (or longer staple wool) would be used for worsted goods. Long-staple wool is more valuable than short-staple, and sheep with long-staple shear greater amounts.

Color and texture also influence the value of the wool; white wool, preferably bright and lustrous, is preferred by the industry. This wool can be dyed any shade or used in its natural color. Black wool is customarily discounted by the mills, but is much in demand by handcraft buyers. Black sheep were sought by the missions for black wool from which to make the monks' habits. Black sheep are not truly black; their wool varies from brown through gray to black and produces attractive varieties of shading in the finished product. In the Shetland Islands, such grayish sheep are known as Moorit sheep, and it is their very fine, soft wool that is used in the beautiful hand-knit sweaters and shawls made in these islands. Other wool is spun into a thread so fine and delicate that shawls are knitted of such lacy texture that they may be drawn through a wedding ring.

Modern technology has made great strides in improving and modifying the processing of wool so that it may now be made machine washable and shrink resistant. New wool yarns are being developed that will have all the desirable qualities of the original fiber with some of the advantages of the synthetics.

Wool is the only natural fiber that has felting and matting qualities. Each strand of wool has on the outer surface, or epidermis, scales that have saw-tooth edges, and it is these

143

scales that cause the felting of pure wool when it is scrubbed during laundry; the scales also give wool the strength to make a durable spinning thread. Wool is very elastic, because of its crimp. Fine wools have more crimp than coarse wools, caused by uneven growth in the cortex of the wool fiber. This characteristic is valuable in spinning as the amount of crimp determines the elasticity of the wool.

Coarse wool has smoother fibers, with fewer scales, and makes a smooth, shiny cloth. The health and well-being of the animal have a significant effect on the luster, or power to reflect light, of its wool.

All wool fibers are poor conductors of heat. Several layers of lightweight woolen clothing are ideal for work under conditions of extreme cold, and are not only warmer than one heavy garment but can be shed one by one as the body temperature rises. Since wool readily absorbs and gives off perspiration, woolen clothes are comfortable in summer too. Because wool holds up to 30 percent moisture without feeling wet it is ideal for people who work under damp conditions. That's why Norwegian fishermen have such handsome hand-knit sweaters that last many, many years, and why merchant seamen wear long, heavy seaboot stockings, knitted from raw wool.

Wool has another characteristic very important in clothing and household use—it is fire resistant. Wool exposed to flame will singe and smell strongly, but will not burst into flame. This is especially important in children's clothing and bedding.

MANAGING SHEEP FOR GOOD WOOL PRODUCTION

Proper feed and plenty of it will keep the flock healthy and vigorous and does much to produce a good clip of wool. Wool is high in protein and it is important to supply enough protein in the feed. Shelter and loafing pens should be as dry and clean as is practicable, and grazing areas should be kept free of burry weeds. The kind of feed and care necessary to produce good lambs also produces good wool, and the greatest growth of wool takes place when plenty of green forage is available.

When sheep are sick or off their feed for any reason, or if their feed is abruptly changed, a weak place in the wool fiber forms at that time. This is called tender wool (or feverbreak) and is discounted by buyers. A good wool grader can tell about when a sheep was feverish or out of condition by looking at the fleece.

Burrs, chaff, seeds, and dirt reduce the value of the fleece. Some foreign matter will inevitably get in, but every effort should be made to keep sheep away from concentrations of plants like burdock and foxtail. Hayracks should be built so that the sheep pull the hay out of them and do not have to stick their heads right inside. This is sometimes easier said than done, because sheep just love to get their heads in and search for the best tidbits. Never throw the hay or feed over the sheeps' backs.

Marking with unsuitable branding liquids such as household paint also spoils the fleece. A satisfactory branding paint may be made as follows:

¼ lb lampblack (or paint pigment)   ¼ pint turpentine
¼ lb flour   3 pints linseed oil
¼ pint pine tar

This amount will brand 350 sheep. Axle grease, colored with paint pigment, is easily prepared and is simply smeared under the ram's brisket. Special scourable sheep-branding paints are sold by livestock supply companies. Metal numbers for marking, and crayons, are available from the same source.

FALLING WOOL

A sheep trailing round the pasture with half its wool gone is indeed a sad sight. Look for the cause, for this condition is not a sign of an attack by hungry moths but a symptom of some aggravating cause or disease.

Various conditions and complaints cause falling wool. One thing not often considered is high fever in pregnant ewes and its effect on the wool producing qualities of the unborn lamb. Excessively high temperatures in the uterus may result in destruction of the wool follicles, resulting in "hairy" lambs. Such lambs are usually small and weak and will never grow a good fleece. Iodine deficiency too will affect lambs, and in extreme cases some are born without any wool at all.

Sheep may lose their wool after a period of high fever, as might be caused by infections following mastitis, retained afterbirth, foot rot, or pneumonia. Stress, difficult birth, and injury may also be followed by fever. The loss is not immediately apparent, and the wool may not start to come off for weeks or even months. Chemicals applied to the skin may also cause the wool to fall off. Pour-on insecticides such as Ruelene 25E or Korlan 24E, used for external parasites, must always be diluted; kerosene, when applied for the eradication of maggots, should be applied to the smallest possible area. Much research has been done to find chemicals that would take advantage of this falling-off of wool following application of certain substances, with a view to eliminating the annual haircutting chore, but so far no entirely satisfactory method has been developed.

Sometimes sheep will chew the wool off each other. While the exact cause is hard to pinpoint, several contributing factors may cause this habit: overcrowding; mineral deficiency; starvation; or heavily wooled udders in the dams that cause the young to get mouthfuls of wool when they try to nurse so that they acquire the habit of eating it.

Injections of vitamin B (iron) for possible anemia and rations with high protein and mineral content seem to help prevent wool eating. Green legume hays are excellent sources of protein, or, when these are not available, linseed meal, soybean meal, or cottonseed meal can be fed. Minerals should be available at all times and salt should be iodized.

SHEARING

Sheep are usually shorn in the spring of the year. In areas where the winter is not too harsh, sheep may be shorn in the autumn, in preparation for lambing time, while in areas where the winter is extremely cold it may not be possible to shear until late May or June. Sheep shearers are becoming a vanishing species in the farm flock areas, and it is wise to start looking for one and to make a reservation early in the year. Local county agents and veterinarians sometimes know of shearers, but if they do not, the best way to find one is to talk to someone who has sheep and find out who has done his flock. The annual fluctuation in the price of wool makes it difficult to state what percentage of the income per sheep is derived from the clip, but a good fleece, carefully shorn, must always be worth more than a dry, burry fleece that is full of second cuts, and a shepherd who can consistently produce top quality fleeces has a good chance of developing a profitable market for his wool.

When the time comes for the shearer's visit, the job will go much more smoothly if a few preparations are made beforehand. A clean place must be provided with space for a shearing canvas, or board, a place for stacking the newly shorn fleeces, and an electric outlet for the clipper. Groups of custom shearers who tour sheep flocks in the spring usually have gasoline-driven shearing machines mounted on trucks.

The sheep to be shorn should be penned the night before and fed only lightly. The fleece must be dry and will shear much better if the animal is warm. If the sheep are wet before the shearer comes, they must be allowed to dry off outside—they will not dry if penned in the barn, as it takes fresh air to do the job. No shearer should be expected to run all over the lot to catch the animals, so some system of close penning should be available during the time he is working. If the sheep have young lambs, these should be separated or provided with an escape gate so that they are not in danger of being trampled. Any sheep which seems to be about to deliver a lamb in the near future should be sheared first and put in a warm, comfortable place.

A good sheep shearer will handle the animals quietly and carefully and will take the fleece off in one piece. (For more about this see the next section *how to shear a sheep*.) The fleece should then be rolled and tied and stored in a dry place. Wool absorbs so much moisture that fleeces will remain damp for a long time if stored in a damp place. Never store fleeces in plastic bags because there is always a certain amount of moisture in the wool, and this must be allowed to evaporate. Use only paper twine for tying, available from livestock supply companies; baler twine will leave jute fibers in the wool.

The shearer will probably appreciate the offer of hot coffee or something cold to drink, depending on the weather, and a comfortable place to sit and eat his lunch.

As the shearer finishes each sheep, look closely at it to see whether the feet need trimming, whether there are any cuts or wounds that need attention, and what kind of wool each animal produces. After each sheep is sheared, it takes only a minute to brush it with the Co-Ral and mineral oil smear or to dust on a handful of Marlate, and the de-ticking operation is done for the year. Korlan and Co-Ral pour-on should not be used until shearing cuts have healed.

How to tie a fleece

1. Lay the fleece clean side down on a board or canvas.
2. Remove any dirty belly wool and the tags and put these in a separate pile.
3. Fold the wool from the legs and neck toward the center of the fleece. Since the shoulders and back of the sheep have the nicest wool, the wool from those areas should form the outermost layer of the package.
4. Kneel so that the fleece is between the knees, and compress the ball of wool as neatly and firmly as possible.
5. Tie the package with two wraps of paper twine like a ribbon on a present. An extra wrap or two around the fleece may be needed to keep the wool together.

Sometimes a sheep will catch the shearer off guard in the middle of the job and run across the floor trailing part of the fleece behind it; a woolbox is a useful thing to have to wrap

How to tie a fleece.

such broken-up fleeces, but don't tie the wool too tight, as there must be room for air to get to it.

If, by some unfortunate mischance, you run out of twine, the sheep's own wool can be made to serve. Lay the fleece down and remove tags and belly wool, as above. Draw out the neck wool, twisting as you do so to make a rope. Quite a long strip can be made in this way. Roll the sides of the fleece into the center, lengthwise, and then roll up the folded fleece from the tail end to the neck end and use the woolen rope as a tie.

No one can learn to shear a sheep from looking at a book—it takes hours of practice on the animal. But the following notes and pictures are intended as a guide to a method that has been proven to work well.

HOW TO SHEAR A SHEEP

After shearing your first sheep, you are likely to come away thinking that a self-shearing sheep would be a splendid goal for the breeder's science. Still, by following the steps outlined here, a smoothly working system can be developed with practice. The steps given below are those used by nearly

149

The tools: oil can, antiseptic powder (Bloodstopper), twine, combs, hoof trimmers, electric clipper.

all shearers and, when mastered, will make it possible to shear a sheep with a minimum of both backache for the shearer and scraped skin for the sheep. The object is to remove the fleece in a single piece, without second cuts and without cutting the skin of the sheep; merely getting it off will result in a second-grade fleece and an unkempt animal.

Make the sheep as comfortable as you can. All strokes should be as long and wide as possible, holding the comb flat on the skin, but if the comb rises off the skin of the sheep, don't go back and make a second cut. Relax and take your time—speed comes with experience. Never kneel on the sheep, and try not to stand on the fleece. Finally, strive never

Ready to go.

A.

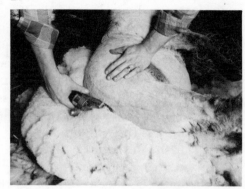

B.

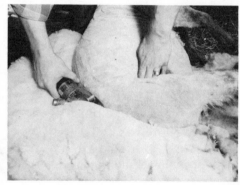

C.

D.

to lose your temper. When the sheep struggles, gets its feet tangled in the electric cord, and then kicks the fleece all over the floor, let it go, have a cup of coffee or whatever will soothe the soul, and then start again.

1. Sit the sheep on its rump with the body between your knees. Holding the sheep's right foreleg across the left side to tighten the skin, make the first stroke down to the left side. The second stroke runs parallel. The third stroke runs down the right side. Remove the remainder of the belly wool by shearing parallel with the first two strokes as far as the middle of the abdomen, then pull the skin towards the brisket and shear from right to left across the stomach, being careful to remove neither teats nor pizzle. On rams, stretch the skin of the scrotum as tightly as possible and shear carefully. Shear the insides of both legs and up as far as the dock with strokes from right to left.

2. Turn the sheep slightly, so that it is resting on its right hip, with the upper half of the body resting against your leg. Pull the skin of the flank towards the head and shear the legs. Then, still stretching the skin of the flank, begin at the dock and shear three or four strokes towards the head.

3. Step between the sheep's legs and, while holding the head with your left hand so that the neck is extended as much as possible, start at the breastbone and shear up to and including the cheek. On the second stroke shear parallel and trim around the horn and the ear. On the third stroke finish behind the ear. Continue until front leg, shoulder, and back of the neck are clear.

4. Step back with your left foot and shear the remaining wool from the left side of the sheep.

5. Roll the sheep onto your foot, holding its head down with your left hand, and take a few more strokes past the backbone to start on the right side.

6. Step over the sheep with your right foot and bring the head back so that it is braced against your knees. Shear down the neck to the shoulder in three strokes, then clear the front leg with strokes from the backbone.

7. Swing your left leg across the sheep to a point even with its hip and roll the sheep over onto its left hip, again resting

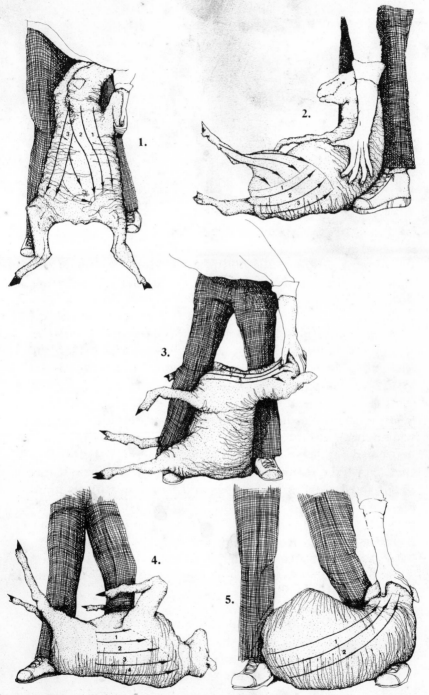

The "tally-hi" shearing method, courtesy *The Shepherd* magazine.

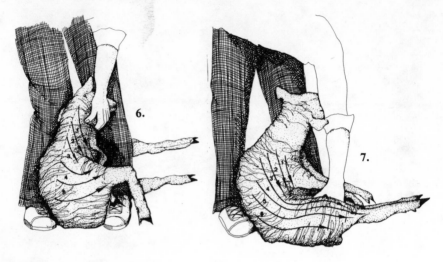

the upper portion of the body against your legs. Shear the right side with strokes from shoulder to flank, continuing down the leg in a smooth line, finishing at the dock. Pictures A, B, and C demonstrate three ways to tighten the skin: A, by holding a handful of skin; B, by pulling the skin taut with the flat of the hand, and C, by using the knuckles pressed in the stifle. This has the advantage of making the sheep straighten its leg and become quieter.

While shearing, do not attempt to hold the forelegs with your hands, and do not allow the sheep's head to drop back between your legs, as it will start to struggle if the head is free. Never hold the hind leg by pulling on the foot. If the sheep starts waving its forelegs and bicycling with its hind legs, put the clipper down and resettle the sheep in a comfortable position. Do not let a helper try to hold either the front or hind legs, because this will just make the sheep become panic-stricken and even harder to hold.

MARKETING WOOL

Wool may be sold to a dealer, to a mill, or through a wool cooperative. In some areas, dealers tour farms and offer to buy the clips from flock owners. Or, the shearers may act as dealers, either buying the wool outright or taking it in part payment for work at the time of shearing. Mills that do custom work may either buy wool outright or take it in part or whole payment for custom-made wool blankets, yarn, or batting. The Shippensburg Woolen Mill in Pennsylvania requires about fifteen pounds of wool to make a blanket, three pounds to make one pound of yarn, and two to three pounds to make one pound of batting (two to four pounds of batting are needed to fill a comfort).

To sell through a cooperative, find out from the local extension service where and when the sale will be held. This may serve one county or a group of counties, and you may

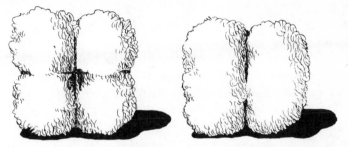

Tie up the fleece so that the wooly side will be inside the bundle.

sell wool at any wool pool, not only the one in your own county. To set the price of the wool, representatives of the wool cooperatives and the agricultural extension service of each state meet with the buyers who intend to bid on the wool. Separate bids are made for the wool from each of the wool pools. When the price is accepted, the representatives of the cooperative notify all prospective shippers of the price to be paid and the date and time of the sale. At the sale, each shipper delivers his wool individually to the grader, who will grade each fleece and then pass each fleece to be weighed. Grades and weights are recorded and a receipt is given. Payment is made in about two weeks.

Some people who are reviving the crafts of spinning and weaving don't have their own sheep to provide the wool, and they are always on the lookout for sources of good-quality fleeces. Reach these buyers by advertising in local newspapers, placing small advertisement cards in craft shops, or visiting folk fairs and craft exhibitions.

WOOL INCENTIVE

In 1954 Congress recognized that wool is an essential commodity that is not produced in sufficient quantity in the United States to meet the country's demand. The National Wool Act authorized incentive payments to producers, financed from funds provided by the collection of duty on imported wool. A small percentage of these payments is withheld to finance promotion and advertising of the lamb and wool industry. This funding of promotion was authorized to continue by a national referendum of wool producers, held in the late 1960s. In order to get the maximum benefit from this program, you should sell the wool for the highest price possible and keep records of such sales. When lambs are sold for butchering, ask the buyer to sign a receipt stating that the lambs are "certified sold unshorn for slaughter," the receipt showing the dale of sale, price, and live weight of the lamb.

Consult the agricultural extension service to learn where to take the receipts.

PREPARING FLEECES FOR HANDSPINNING

Hand spinners need the best possible fleeces—free from seeds, burrs, tags, and manure. The preparation of such fleeces offers a chance to take advantage of a market for good-quality wool that is in top condition and properly handled. Fleeces should be skirted (trimmed) after shearing, and the less-desirable wool removed. Second cuts, cotted (matted) wool, and hardened grease will spoil fleeces for handspinning.

To present the most attractive fleece, find a large, shallow carton and line it with paper or a piece of an old sheet. Remove as much as possible of the foreign matter in the fleece—straw, burrs, and manure—and lay the fleece flesh side up in the box, pushing it together to make it as compact as possible. Put another piece of sheeting or paper over this wool and repeat the process until the box is full. Do not pack the box too tightly: the wool must breathe, and the fleeces will not look so attractive if they are squashed. Label the box with the weight and color of the wool, the kind of sheep from which the wool came, and of course the price. If the buyer wants one fleece from the box, this can easily be weighed by picking up the four corners of the piece of sheeting, pinning them together, and hanging them on the scale.

Black fleeces are becoming increasingly popular and will often command a higher price than white ones. The fleeces can be prepared and stored in the same way as the above, but care must be taken to keep the coarser wool separated from the longer, finer wool. It is a good idea, too, to mark each fleece with the name of the sheep from which it was sheared in case the spinner comes back next year for more of the same kind of wool.

DYEING WOOL WITH NATURAL DYES

The plants from which natural dyes are made are growing all around us, free for the gathering. Early settlers in this country made dyes from plants which were familiar to them, and cooked up the solutions in iron pots that had a graying effect on the colors, the iron acting as the mordant (a mordant is a substance that sets the color in the wool and enables the wool to retain the color after washing). Later, as people experimented with native plants and as chemical mordants became available, the colors became clearer and brighter. Chemicals for mordants are available from paint and hardware stores, or from chemical supply companies.

Indian dyers used to add the mordant to the dye solution, thus doing the dyeing and mordanting all at one time. Some types of mordants are added to the dye after the dyeing is completed and before the wool comes out of the dyebath; but for best results, the wool is mordanted first and then dyed. Alum or alum and cream of tartar are the easiest and safest substances to use; other mordants are chrome (potassium dichromate), copperas (ferrous sulfate), and tannin. Common salt and vinegar are also mordants.

157

A variety of shades may be obtained from a single dye by varying the mordants used. For instance, cochineal mordanted with alum gives a rose color, with cream of tartar, bright red, and with chrome, purple.

The equipment needed for dyeing wool is simple:

- Enamel kettles or tubs to hold the dye. Metal pots will change the color of the dye.
- Tubs for rinsing.
- Scales for measuring mordant that will weigh in fractions of ounces.
- Measuring cups.
- Cheesecloth to strain the dye.
- Wooden spoons with long handles for stirring the dye and turning the wool. If the pot is very deep, dowel rods, sanded smooth, will serve.
- Thermometer, rubber gloves, soap.

Mordanted wool may be dried and stored until needed for dyeing. To mordant a batch of wool, first wash it (see the following section, *washing wool*) and put into the solution when wet. Quantities of chemicals vary, depending on whose recipe is used. Rita J. Adrosko, whose book *Natural Dyes and Home Dyeing* (Dover, New York, 1971), is an excellent source of information for dyeing enthusiasts, recommends using 4 ounces alum and 1 ounce cream of tartar for 1 pound of dry wool, dissolved in 4 or 4½ gallons of water. The alum enables the wool to absorb the dye, and the cream of tartar sets the dye so that it may be washed without fading.

Dissolve the alum and cream of tartar in cold water. Immerse the wet wool. Gradually heat the solution almost to boiling (210°F.) and simmer at this temperature for about one hour; gently turn the wool over several times to allow the mordant to soak in evenly, but do not stir, as this will cause the wool to become tangled. At the end of an hour, turn off the heat and leave the wool in the solution to cool. When the wool is cool it can be removed to an enamel colander to drain, pressing out excess moisture. Wrap the wool in a towel to remove as much of the remaining moisture as possible, and then wrap in a dry towel and store in a cool, dry place.

For chrome mordant, dissolve ½ ounce of potassium dichromate in about 4 gallons of cold water. A traditional mordant is made of vinegar and salt, the proportions best determined through experimentation. For a dye from the moss that grows on oak trees, add 5 cupfuls of salt for every 2 gallons of moss and water.

WASHING WOOL

Dye solutions will penetrate the wool more thoroughly and evenly if the wool is washed before it is dyed. If the wool is not yet spun, carefully pull apart the fleece until it is soft and springy and place it in a cheesecloth bag. Dye and mordant recipes are based on dry weight, so weigh the wool before washing, then immerse it carefully in warm, soapy water. If the water is very hard, use a water softener, or add some vinegar. Squeeze gently, since rubbing or twisting will mat the wool.

Rinse in soft water of the same temperature as the wash water, using two or more rinses. When the water runs clear, squeeze out excess water and set the wool aside until ready to use. If it is not to be used at once, put the wool on a towel and lay the towel over a drying rack. When dry, roll the fleece in a clean towel. Don't leave damp fleeces lying around too long, as they may mildew.

THE PROCEDURE

Use clean wool, and be sure that the fibers are well wetted before putting them in the just-simmering dyebath. Wool should not boil hard, nor should it cook too long. Keep the wool under while in the dyebath. Do not stir, as this will make the wool tangled, but turn the wool gently with a wooden paddle or spoon. After ten minutes, remove the wool and add 1 tablespoon of Glauber's salt to encourage uniform dyeing, and return the wool and simmer thirty minutes. In deciding when to take the wool out, allow for the color lightening as it dries.

Dye all the batch of wool needed for one job at one time, as it is almost impossible to duplicate any shade exactly.

If the color dries unevenly, correct by returning the wool to the dyebath, to which Glauber's salt has been added in a quantity equal to 40 percent of the weight of the dry yarn.

Rinse the wool thoroughly after removing it from the dyebath, and hang in the sun to dry. Some dyers feel that hanging the wool in the sun brightens the colors, while others prefer hanging in the shade.

Dyestuffs and mordants are often poisonous and some irritate the skin, so they should be kept out of reach of children. Wear rubber gloves. Never wring or twist wool, as this will cause streaking of the dye and matting of the fibers. Changes in the temperature of the water cause shrinkage, so after the dyeing process is completed, allow the water to cool gradually before changing to rinse water; keep all rinse waters at the same temperature.

The materials listed here are easily found and harvested; many of them can be dried and preserved until you are ready to use them. Leaves and roots of a plant should be used before the plant comes into flower. Flowers should be gathered at the end of summer, after long days of sunlight. Bark is best in the spring when it has most sap. Oak galls, walnut hulls, and the like are harvested in the fall, and berries and fruits should be fully ripe.

Exact shades cannot be described here, because growing conditions affect colors, and methods of dyeing lighten or darken shades. Half the fun of dyeing is in experimenting with different flowers and substances.

• *Sumac.* One gallon of ripe sumac berries plus 1 teaspoon of copperas will dye 1 pound of wool dark gray. Soak the berries in water for an hour, then boil for thirty minutes and strain. Return the liquid to the kettle, add water to make about 4½ gallons of dye solution. Add copperas and mix well. When the dyebath is hot (150°F.), add the wool and bring water to simmering (190°F.). Simmer for thirty minutes. Cool and rinse.

• *Acorns.* Seven pounds of acorns plus 2 ounces of alum will give a tan color. Put the acorns in the kettle and boil for

2½ hours. Strain the liquid, then add water to make about 4½ gallons. Add alum, stir until thoroughly dissolved, then bring the solution to 150°F. Immerse the washed wool and bring to 190°F. for thirty minutes. Cool and rinse. For a grayish color, use copperas.

• *Onionskins.* Crush 8 ounces of onionskins and soak in water for at least one day. Bring the solution to a boil and simmer for one hour, then strain the liquid and let it cool. Add the mordant (4 ounces alum and 1 ounce cream of tartar for yellow, or 2 ounces chrome and 1 ounce cream of tartar for deep gold), heat to 150°F., and add wet wool; then heat to 190°F. and simmer until the wool is several shades darker than you want—it will lighten as it dries. Let the wool cool in the dye, and rinse and dry in a shady place.

• *Lichens.* Lichens are mosslike plants that grow on rocks, trees, old buildings, and old fences, and can be found almost anywhere. They are most easily gathered in wet weather, when they are the most vital. Lichens usually need no mordant, although a sample of the finished product should be sun-tested for several days to see if it fades. With the addition of alum as a mordant, varying degrees of yellow-tan to dark rose are the result. Lichens are pleasant smelling and make the wool soft to the touch.

For 1 pound of wool use 1 pound of lichens. Break up the lichens and put in a kettle, cover with water, and boil one hour. Cool the dye, add wet wool, and simmer until the dye is the desired color. If a mordant is to be used, add this to the dye solution, using 4 teaspoons copper sulfate and 1 teaspoon cream of tartar. When the wool is the right color, cool and rinse, then shake out the pieces of lichen.

• *Nettles.* The roots of the stinging nettle, boiled with salt or alum, produce a yellow dye.

WASHING WOOL SPUN IN THE GREASE

Fill a tub with warm, softened water and add enough mild soap to make suds. Tie the skein loosely in several places—if the wool is loose it must be wound on a skein as loose wool will tangle—and immerse it in the water. Leave it there for about twenty minutes and then squeeze gently, but do not twist. Remove from the suds and squeeze again. Rinse in softened water of the same temperature until the water runs clear. Hang the skein over a rustproof peg or hook and attach a weight to stretch the yarn: two potatoes tied in cheesecloth, with the ends of the cheesecloth tied round the wool, make a good weight. When the wool is dry it can be coiled and stored in a mothproof bag.

The amount of soap used will determine how much grease is left in the wool. For seaboot stockings or fishermen's sweaters, some natural grease should be left in the wool. Wool that is to be dyed must always be rinsed free of grease.

Harrisville loom, made by Harrisville Designs, Harrisville, N.H.; Ashford wheel, sold throughout the United States; Lazy Kate for holding extra bobbins; cards for preparing wool for spinning.

SPINNING FROM WOOL
IN THE GREASE, WITHOUT CARDING

Choose a clean fleece with long staple, free from burrs and not cotted (matted), and place it sheared side down on a table. Pick up a small bunch of wool and comb it, with a metal-tooth dog comb, from the skin end to the tip until it is clean and soft and smooth. Do this until you have enough wool to cover a large tray, or board. When there is no more room, place a sheet of paper over the combed wool and make another layer. Don't make more than two or three layers, or the weight of the wool will squash the bottom layer. Hold the wool loosely while combing so that the fibers don't mat, and if the wool is very dirty, comb both ends of the lock of wool. Keep the wool in a warm place; if it has become very hard to work, put it in the oven for a few minutes to soften the grease.

CARDING WOOL

When any substantial quantity of wool is to be spun, it is customary to first card the fleece. Carding is the process of separating the fleece and preparing it for spinning, using wool cards—small boards to which is affixed a piece of cloth studded with small hooks. These cards come in a variety of sizes for trimming show sheep, but one pair of No. 8 cards will serve very well for carding wool that is to be spun.

To use the cards, take a small tuft of wool, about half a handful, and spread this over one of the cards. For a right-handed spinner, take the bottom card in the left hand and hold it facing inwards and with the handle uppermost; with the right hand, draw the card, which is loaded with wool, across the other card. Do this two or three times, changing the wool from one card to the other, until the wool is soft and fluffy. Then make a "batt" of the wool by drawing the top card once forward and once or twice backward, so that all the wool ends up on one card. Next, pull this batt apart very slightly, and then roll the wool *very lightly* between the palms to make a roll of wool.

SPINNING

Spinning with a wheel or with a handspindle takes lots of practice, but the directions here are enough to get you started. To start spinning with a handspindle, take a piece of ordinary knitting wool and wind it a few times round the base of the spindle. Next, take the thread down and wind it once around the knob under the base of the spindle. Then bring it up and wind it once around the top hitch of the spindle and pass the thread under the loop, drawing it up firmly but not tightly. Pick up a lock of wool and, while holding the tips of the lock of wool against the thread, give the spindle a twist to start it spinning. Draft the fanned tips of the wool against the thread through thumb and forefinger, keeping the thickness of the newly spun wool as even as possible. When the thread has become so long that the

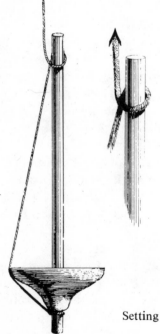

Setting up the handspindle.

spindle is too far down to reach it, take the loop off the top hitch, wind the wool on the spindle, and then start again. This is a good way to get the feel of wool and spinning. The spindle need not be expensive—in fact you can make one with a thick knitting needle and a cake of soap.

SPINNING WITH A WHEEL

Spinning wheels come in a variety of shapes and sizes, but they all work on the same principle. Before starting to spin, study the wheel and learn how it works. Practice pedaling, making sure the wheel turns to the right, and maintain the rhythm of the pedaling while doing something else with your hands. The way you sit will depend on the type of wheel you are using, but in any event, be comfortable and relaxed.

To start spinning, take a length of knitting wool, tie it to the bobbin and pass the end over the hooks of the flyer and through the eye of the spindle. Take a lock of the unspun wool in the left hand and lay the thread on this. With the right hand start the wheel, and as the twist runs up the thread, draft the unspun wool between the thumb and forefinger of the left hand onto the thread. Keep the fingers moving to draft the wool evenly, and keep the pedal moving rhythmically. If the thread keeps breaking, the tension may be too tight; if the thread is knotted, the tension may be too loose. It takes a lot of practice to keep everything moving smoothly and evenly.

Keep the unspun wool at your left knee so that there is a minimum of movement required to keep a supply of wool in the fingers. If the wool twists too much and becomes knotted, it may be that the driving belt needs tightening or that you are pedaling too fast. If the wool is not twisting enough, the tension is too tight.

As the bobbin fills, unhook the wool and move it to the next hook so that the bobbin fills evenly. When the bobbin is full the wool must be wound on a skein winder before washing. Any board with two pegs about 18 inches apart (or a handy helper with two willing hands) can be used for winding wool into a skein. When the wool is wound, tie it in four places and then wash.

WEAVING

Having spun the wool, and possibly dyed it, the next step is to weave the yarn into cloth. Clothes have been woven since the late Stone Age. In ancient Greece and Rome the weaving of silk, wool, linen, and cotton cloth was highly developed. Today, in all parts of the world, weaving is done both in primitive and highly technical societies; the main difference between the methods used is merely a degree of sophistication in the tools employed: from fingers to multigeared automated looms, and every variation in between.

For a beginner, the backstrap loom is a good instrument to start on. It produces a plain weave of any length, is easy to use, and costs relatively little. You can use yarn that has already been dyed and plan your pattern with it, or it is possible to dye a finished piece. However, the latter method tends to lead to a few problems, such as finding a pot large enough to prevent foldovers, being aware of possible mottling occurring, and being aware of shrinkage. To dye the finished product involves the same procedure as dyeing yarn.

Yarn for weaving can be spun with other fibers such as dog hair or milkweed seeds added for varied textures. When dyed, this mixed yarn will have varied shades and colors due to the ways the different materials take the dye.

TYPES OF LOOMS

Beka and Leclerc both make excellent backstrap looms for a moderate price. It is better to experiment with a small model and move up than to invest larger sums and become disillusioned or frustrated. Sunset and Step-by-Step books are informative and inexpensive, and the Handweavers' Guild of America publishes a quarterly magazine, *Shuttle, Spindle, and Dyepot,* that contains invaluable information on spinning, dyeing, and weaving for beginners and experts.

The inkle loom is used for weaving narrow bands, which may or may not be sewed together for added width. Used mainly for belts, this loom produces a warp pattern. The card loom is not a loom per se, but rather a type of weaving

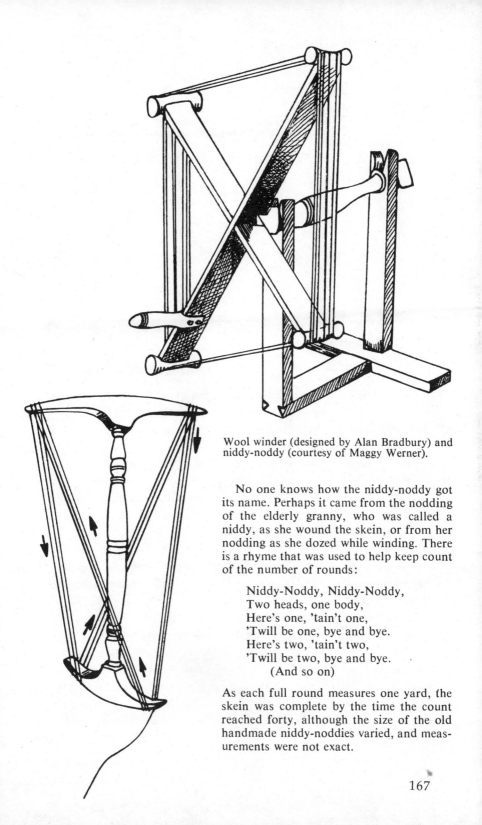

Wool winder (designed by Alan Bradbury) and niddy-noddy (courtesy of Maggy Werner).

No one knows how the niddy-noddy got its name. Perhaps it came from the nodding of the elderly granny, who was called a niddy, as she wound the skein, or from her nodding as she dozed while winding. There is a rhyme that was used to help keep count of the number of rounds:

> Niddy-Noddy, Niddy-Noddy,
> Two heads, one body,
> Here's one, 'tain't one,
> 'Twill be one, bye and bye.
> Here's two, 'tain't two,
> 'Twill be two, bye and bye.
> (And so on)

As each full round measures one yard, the skein was complete by the time the count reached forty, although the size of the old handmade niddy-noddies varied, and measurements were not exact.

167

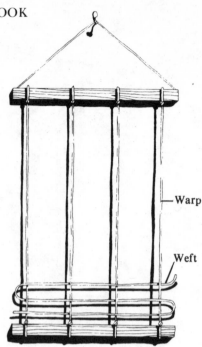

Warp

Weft

A beginner's loom.

used for bands. These bands have more strength than inkle bands because the warp is twisted together as it is woven. This method produces a warp pattern, and allows for greater pattern variation.

This backstrap loom can produce a piece as wide as the rigid heddle used or as narrow as you wish. Because of the versatility in width and length of the piece and also because of the small storage space needed (about 2 x 8 x 30 inches) when not in use, this is a good loom for the beginner.

A beginner's loom can be made with a frame of any desired shape or size that is convenient to handle. It is possible to insert the weft between the warp with either a shuttle or the fingers.

Harness looms are floor looms, and usually come in two, four, eight, or twelve harnesses. Each harness holds particular warp threads and allows more variations plus speed in weaving. Traditional coverlets were woven on four- or eight-harness looms. Besides the added freedom of pattern, the standard floor loom also allows greater yardage to be woven on one threading and larger widths. This type of loom produces a weft pattern.

SHEEPSKINS

By the end of the summer, when the lambs have reached market weight, they will be carrying a fleece of healthy wool, two or more inches long. When these lambs are butchered, ask the butcher to salt the hides immediately, or to put them in the refrigerator. As soon as possible, pick up the hides and get them to a tannery. Some butchers are reluctant to keep the hides on the premises because of local regulations, and it is worthwhile going back the same day the lambs are butchered to pick up the skins. When you do this, salt them as described in the following section.

Tanned sheepskins will be white, gray, or any other shade depending on the kind of wool the sheep carried. Use them for car seats, rugs, pillows, sheepskin coats—their uses are endless. For making sheepskin jackets, it's best to use the hides off spring lambs shorn about two months before slaughter; these skins will be soft and carry about ¾ inch of wool. Toys, mittens, and pillows are best made with the softest and youngest lambskins available. Moccasins and car seat covers will be better with the heavier hides. Heavy sheepskin vests, of the sleeveless hip-length variety, are made with fleeces of full-grown sheep. Special needles with flattened points should be used. These are made by Risden and can be obtained from large fabric centers and some craft shops. Button thread, crochet thread, or waxed dental floss will all make strong seams.

The comparatively new glutaraldehyde tanning process that makes sheepskins machine washable is recommended for hides to be made into toys for children, bedpads for invalids, or outer garments that may get wet. Such bedpads are invaluable for bedridden patients who may have a tendency to soreness or bed sores, and will also be a great comfort to those who feel the draft a lot.

TANNING

Leftover pieces from the making of toys or articles of clothing can be trimmed and used to make an afghan. If the stitching is strong and a lining of velour or some other soft, fairly heavy material is sewn on the back, such an afghan will last a lifetime.

Because there are several different formulas for making a tanning solution, it is not possible to give any one reliable method for washing or cleaning sheepskins. One method that can be used on any kind of sheepskin is to wet some sawdust or cornstarch very slightly with dry-cleaning solvent and very lightly rub this into the wool side of the sheepskin. Then shake thoroughly and brush up the wool with a wide bristle brush, or fluff it with a wide-toothed comb (the kind that is sold for shaggy dogs). Skins tanned with glutaraldehyde will have a slightly golden tinge to the skin, and can be washed easily in cold water and Woolite, and then rolled in thick towels to take up the excess moisture. They should be hung to dry, away from heat. If necessary, invalid washable sheepskins can be washed according to hospital sterilizing standards. We have successfully washed alum-tanned sheepskins in Woolite, and also in a solution made with

---

Preparing Skins for Shipment to a Tannery

Skins to be shipped raw for tanning should be:

• *Free from cuts.* One or two small cuts (one inch or two inches) are alright if around the edges, but not if in the middle.

• *Salted.* Three to four pounds of granulated salt (*not* rock salt) should be enough for two large hides, when evenly distributed.

• *Dried.* Three-quarters dry is ideal—a little stiff, but still flexible. Four or five days out of sun and rain should do it.

Plastic garbage bags tend to sweat inside in warm weather, so they are not a good idea. A feed bag with skins inside, packed in a cardboard carton, is the best arrangement.

---

enough soap flakes to make a tub of suds, plus 8 ounces of methanol and 2 ounces of oil of eucalyptus. The skin should be left to soak for a few minutes and then swished around in the suds. Do not rinse the skin. Squeeze out excess moisture and dry away from direct heat.

The processing of animal skins to make them soft and wearable is almost as old as the domestication of sheep and cattle. We have come a long way since the days when the animal was skinned and the skin slung into a urine pit to cure for a few weeks. Today's tanneries use a variety of solutions and processes. One of the earlier methods was soaking the skins in tannin, a vegetable extract of the bark of such trees as hemlock and oak. Bark-tanned skins have a brownish color: white oak bark gives a yellowish shade, and chestnut gives a dark brown color. Salt and alum were used in some places. The green, or raw, hide was salted down on the flesh side after all the flesh and fat had been scraped off. When the hide had absorbed the salt, the excess was shaken off and the hide covered with alum. In about a week or so the hide was thoroughly dry and ready for use. The skin could be made soft and pliable by working it between the hands.

Alum tanning is the most reliable of the tanning methods that can be used on a farm; it is safe, simple, and does not require a large investment in equipment. The skins of the butchered animals are liberally salted and placed on wooden slats or skids to dry. If enough salt is used—at least one pound per hide—and if the skins are allowed to drain and dry, they will keep several months this way. When it is time to tan them you will need a large tub or stone crock. Soak each skin for two hours in lukewarm water to soften it, then rinse off with several changes of water, until the water runs clean. This will wash out bits of dirt and manure, and the skin can then be washed in any laundry soap or detergent. Scrape off fat or flesh during this washing, and rinse in clean water. Extract as much water as possible, either by squeezing in towels or by putting the hide through a wash wringer.

Next comes the alum and salt. The hide, fleece side down, may be rubbed with a mixture of 4 ounces of alum and 8 ounces of salt, or soaked in a solution of 4 ounces alum and

2 ounces salt per gallon of water. Allow the hide to soak for three days, turning twice a day. The Vermont Sheep-breeder's Association recommends using a solution made of 1 pound alum and 2 pounds salt and water added until the skin is completely covered. If the alum and salt are rubbed in, the pelt is left overnight with the mixture on it and next morning is hung over a rail, skin side out. Next day the skin is sponged off to remove the mixture and a few ounces of glycerin or neatsfoot oil rubbed in to soften the hide. As the skin dries it will tend to become hard and must be worked to soften it. This can be done by drawing the skin back and forth over the blade of a scythe or over the edge of a table.

Once the pelt is dry and softened, it should be sanded. Those handy little sanders that are sold for cleaning eggs work very well; or, an electric sander may be used.

Other substances used for tanning include Tannol, Lutan, and a combination of Lutan and Chromitan. Methods for using these substances are outlined in *Handbook of Information for the Vermont Sheepbreeder*, issued by the Vermont Extension Service (Publications Office, Morrill Hall, Burlington, Vermont 05401).

HOME TANNING WITH GLUTARALDEHYDE

Woolskins tanned with glutaraldehyde can be washed, an advantage over more traditional methods. The following description of this tanning process is by William F. Happich of the Eastern Regional Research Center, Agricultural Research Service, USDA, and is reproduced here by permission of *The Shepherd* magazine. The photographs were also supplied by the Research Center.

There are many purposes for which launderable, home-tanned woolskins can be used—for example, as pads for tractor, truck and chair seats, and as throw rugs. The wool can be sheared evenly and handsewn articles can be made. The optimum wool length depends upon the use of the woolskin. For example, bedpads may have a sheared-wool length of one inch, rugs two to three inches. When the wool length is four inches or more and is felted and matted or starting to shed, the skin is not suitable for tanning as a woolskin. However, since the finishing operations to process the wool-skins for use as bedpads or paint rollers require special machinery and skilled operators, the home-tanned-and-finished product may not be suitable for these uses.

Numerous requests have been received from sheep breed-er associations, sheep ranch owners, farmers, and others for information concerning the home tanning of woolskins with glutaraldehyde. Accordingly, a simplified procedure was developed that could be used by the individual sheep raiser and requires only a minimum of equipment and chemicals.

• *Preparing for tanning.* Skin the sheep immediately after a slaughter. Be careful to avoid cutting or scoring the skin, as this will damage the usefulness. Remove the head and ears (illus. A). Cut off the excess flesh and fat.

Scour by stirring the skin for about five minutes with a wooden paddle in a solution containing 1 to 2 cups of a mild soap or detergent dissolved in approximately 11 gallons of water at a temperature not over 90°F. (B). Drain over a board or sawhorse, then rinse in several changes of water at the same temperature. The scouring and rinsing should be repeated until the skin and wool are clean.

Place the skin, wool side down, over a wooden beam. Trim the ends of the legs and scrape or cut off any remaining flesh (C). Then scrape the entire flesh side with a tanner's fleshing knife or with a butcher knife held firmly against the skin, pushing away from the body (D). Use only moderate pressure to avoid cutting holes in the skin and remove the layer of fat, elastic, and muscle tissue. Rinse the flesh side with water. Drain, flesh side up, over a wooden sawhorse or board until the water has drained off, for twenty to thirty minutes. Then squeeze the excess water out of the wool and skin by hand and weigh. Record the weight in pounds and ounces. Refrigerate or cool the skin overnight if possible but do not allow it to freeze. If tanning cannot be started the next day or if there are more skins than can be handled at one time, the flesh side of the skins should be rubbed with a substantial amount of fine-grained salt, then covered with additional salt. Fold in the flesh side from each end and store

A. Removing the head and ears from the woolskin.

B. Scouring the woolskin with soap or detergent.

C. Trimming the woolskin.

D. Scraping the flesh side of the woolskin with a butcher knife inserted in a block of wood.

in a cool location. Before tanning, soak the skin in cool water overnight.

• *Glutaraldehyde tannage.* For each pound of the drained, wet weight of the scoured woolskin, place 5 quarts of water (approximately 85°F.) in a clean, watertight wooden barrel (E). Stainless steel kettles are satisfactory, but iron or galvanized tubs should never be used for tanning. Add ½ pound of technical grade salt for each gallon of water and dissolve by stirring with a wooden paddle. Measure 2¼ fluid ounces of glutaraldehyde (25-percent commercial solution) for each pound of the drained, wet weight of the scoured woolskin, pour it carefully into the salt solution, and stir well (E).

Glutaraldehyde is irritating and contact with the skin and eyes or inhalation of vapors should be avoided. It is wise to use rubber gloves, a rubber apron, and a safety-visor or safety glasses; adequate ventilation is important, also.

The following is an example of the chemicals that were used to tan a woolskin in the laboratory.

	%	Weight	Volume
Woolskin, scoured, wool length 2.5 inches	100	13 lb 15 oz	
Water, approx. 85°F	1,045	145 lb 11 oz	17½ gal
Salt, technical grade, percent on solution basis	6	8 lb 12 oz	
Glutaraldehyde (25% commercial solution)	15	2 lb 2 oz	1 qt

(The water should weigh 10½ times the weight of the skin. The weight of the salt should be 6% of the weight of the water. The weight of the glutaraldehyde solution should be 15% of the weight of the water.)

Immerse the woolskin carefully in the glutaraldehyde solution to avoid splashing. Stir for about five minutes with a wooden paddle, then for one minute at hourly intervals during the day. Cover the barrel with a wooden cover between stirrings and overnight. After several hours the color of the wool and skin becomes pale yellow as tanning proceeds. Allow to stand overnight with the woolskin completely im-

E. Preparing the glutaraldehyde solution.

F. Cutting a sample of woolskin for the shrinkage temperature test.

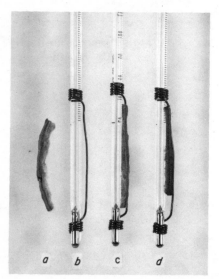

G. Shrinkage temperature apparatus (b.) with sample of skin before (a., c.) and after (d.) shrinkage.

H. Fat-liquoring by rubbing in the oil emulsion on the flesh side of the woolskin.

mersed. Stir one minute per hour the second day. Continue the tanning for at least forty-eight hours.

Progress of the tannage may be followed by determining the shrinkage temperature of the skin as follows. Cut a ¼" x 2½" piece, preferably from the neck area (F). Trim off the wool. Attach one end of the skin with a copper wire guide to the bulb end of a thermometer, allowing the lower end of the skin to hang free (G). The guide can be made easily from #16 or #18 soft copper wire. The skin should be held with sufficient pressure so that the free end does not float out into the water, but not so tightly that it cannot shrink. Hold the sample of skin in a pan of water at room temperature, then heat slowly. Note the temperature at which the skin starts to shrink. This is known as the shrinkage temperature or Ts. When the skin has a Ts of at least 176° to 185°F. (80° to 85°C.), the glutaraldehyde tannage has been completed. Even though the maximum Ts can be obtained in about forty-eight hours when the tanning solution is about 70°F., the glutaraldehyde continues to combine with the skin and a fuller, softer leather can be obtained by continuing the tanning for another six or eight hours. Drain, then wash in several changes of water. Hang the skin, wool side up, over a wooden sawhorse or board overnight.

• *Fat-liquoring, drying, and staking.* Weigh accurately an amount of a suitable fat liquor oil (for example, a highly sulfated neatsfoot oil or a sulfited sperm oil) equal to 2 to 3 percent of the original drained weight of the scoured woolskin. If the skin is very greasy, reduce the amount of fat liquor. Add an approximately equal volume of warm water while stirring to form an emulsion. Add a volume of clear, household ammonia equal to ½ percent of the drained weight of the scoured woolskin and stir. Divide the fat liquor into two equal volumes. Place the woolskin on a flat surface. Apply a small amount of the fat liquor emulsion by hand or bristle brush to an approximately ten-inch square area of the flesh surface and spread evenly and quickly with a back-and-forth motion (H). Repeat until the entire flesh side has been covered. After thirty minutes apply the second portion of the fat liquor as above. With practice the fat liquor can be applied uniformly. Place the skin, flesh side up, on a flat surface and cover overnight with an impervious sheet of plastic such as polyethylene. If several skins are fat-liquored at one time, they may be piled overnight, flesh side to flesh side.

The next day place the skin, wool side up, over a wooden sawhorse and dry the wool at room temperature. An electric fan may be used to speed the drying. Then nail the skin, wool side down, to a plywood board, stretching it slightly. Space the nails (#6 finish) every five or six inches around the circumference and about ½ inch in from the edge (I). Dry the flesh side at room temperature. When nearly dry, test the edge of the leather by stretching a small area with the fingers. When the color lightens considerably, pull the nails and stake the leather by pulling the flesh side repeatedly back and forth in all directions over the edge of a staking board (J). The top of the board is rounded and beveled to a $1/16$ inch edge. Staking must be started at the proper moisture content. Continue staking at intervals until the leather is dry. If the dry leather is not soft and pliable, dampen the flesh side lightly with water and cover with plastic for at least several hours, or preferably overnight. Dry and stake as above. If necessary repeat the dampening and staking until the leather remains soft and flexible.

Grease spots on the dry leather can be removed by sponging with a dry-cleaning solvent or by dipping in white gas. Solvents should be used in a well-ventilated area.

• *Finishing*. The flesh side of the leather may be cleaned and made smooth by rubbing lightly with medium-grade sandpaper fastened to a block of wood (K), or by use of a sanding machine. The thicker areas such as the neck may be sanded to make the leather thinner and more flexible.

The wool should be combed out carefully, using a wide-spaced metal-tooth dog comb (L). The narrow-spaced side of the comb may be used for the second combing. If desired, the wool may then be combed to a vertical position and sheared evenly.

Woolskins or shearlings tanned with glutaraldehyde can be successfully laundered many times by hand or in an automatic washing machine. Use moderately warm water at not over 120°F. (fine fabric setting) and a mild soap or detergent. Wash for five minutes. Longer washing periods may mat or

I. Drying the woolskin on a plywood board.

J. Staking the leather on a staking board.

K. Sanding the flesh side of the woolskin.

L. Combing the wool — the final stage of the process.

179

felt the wool. Rinse with clear, cool water several times. If necessary, repeat the washing and rinsing. Spin to extract the water. If hand laundered, shake out the excess water by hand. Hang at room temperature to dry, which will require about forty-eight hours. If desired, the wool may be combed with a metal-tooth dog comb.

• *Material Suppliers.* Fat liquor oils may be obtained from the Salem Oil and Grease Co., 60 Grove Street, Salem, Massachusetts 01970; Nopco Chemical Co., 60 Park Place, Newark, New Jersey 07102; Reilly, Whiteman, Walton Co., Conshohocken, Pennsylvania 19428; or other specialists in tannery oils.

Glutaraldehyde (25 percent commercial solution) may be obtained from the Union Carbide Chemical Co., 270 Park Avenue, New York, New York 11205.

(Mention of brand or firm names does not constitute an endorsement by the USDA over others of a similar nature not mentioned. Photos by Clifton Audsley.)

THINGS TO MAKE FROM LAMBSKIN

• *Lion pillow.* Use baby lambskin (brown or gray shaded is the most suitable) and cut two ovals as large as the skin will allow. With a kitchen saucer as a pattern, cut a circle out of one skin, as shown. On a piece of tan felt or other soft material, draw a lion face, and embroider with thick wool. Sew this face onto the piece of skin with the hole in it, then sew the two pieces of skin together, leaving a small hole through which you stuff the pillow with old nylons or washable foam. Then, sew up the hole. Surround the face with a border of fuzzy wool or a band of darker sheepskin.

• *Cat pillow.* Cut two circles from black or gray lambskin, and embroider the face with a double strand of very thick, soft wool. Sew all around except for a small hole, and stuff. Add two ears.

• *Baby doll.* Use two pieces of white lambskin, or washable lambskin for front and back. Cut a circle out of the top of one piece. Draw the face on a piece of soft white material— an old undershirt is good—and embroider in pink and blue wool, with a brown curl for hair. Sew the face into the skin. Sew the two pieces of skin together, leaving a small hole, and stuff the doll. If the doll is for a very small baby, do not stuff it—the toy will be better if soft and flat.

• *Shoe buffer.* Sew a strip of lambskin or sheepskin onto a strip of leather or heavy cloth.

• *Dust mop.* Sew scraps of long-wooled sheepskin around the top of a dowel rod. At the bottom of the mop, wrap some adhesive tape once or twice round the rod, and anchor the mop to this with needle and thread.

• *Hunter's mitt.* Sew an oblong piece of lambskin on three sides to make a pocket, and attach a loop to one corner so that the mitt can hang on the belt. Use with the wool on the inside.

• *Moccasins.* Use a piece of lambskin about ten inches wide for a woman, twelve inches wide for a man, and four inches longer than the foot. Cut two slits in the back (illus. 1), and discard piece A (2). Sew B to C. Turn up piece D and stitch all round (3). Sew up the front. Turn down the top and stitch all round (4), and thread a lace through this tube.

• *Mittens.* Save all discarded pieces of sheep or lambskin; they are great for polishing woodwork, shining up the family car, lining winter boots or shoes, padding heavy leather shoulder straps, and for bicycle seats. Pieces too small to be used for anything else can be put in a shallow box to make a luxurious bed for the family cat.

Hunter's mitt.

Cat pillow.

Shoe buffer.

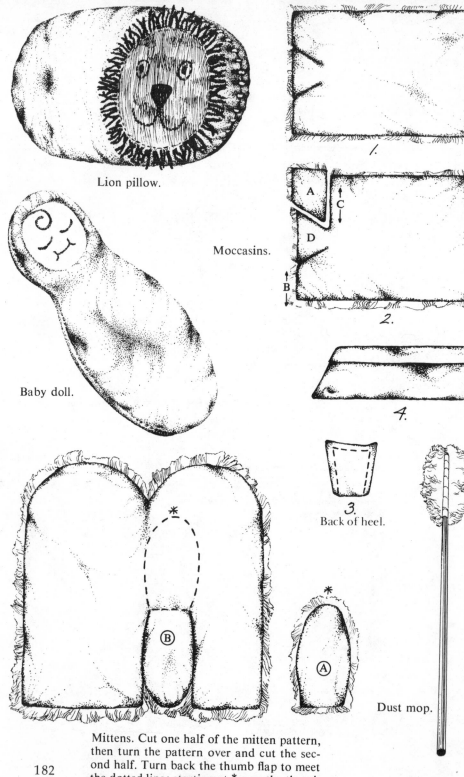

Lion pillow.

Moccasins.

Baby doll.

A

C

D

B

1.

2.

4.

3.

Back of heel.

Mittens. Cut one half of the mitten pattern, then turn the pattern over and cut the second half. Turn back the thumb flap to meet the dotted line; starting at *, sew the thumb piece (A) to the thumb flap (B).

Dust mop.

182

8

Fitting
and
Showing
Sheep

Fairs and livestock shows are the stockman's show window. They provide an opportunity for breeders to come together and exchange ideas, to compare their stock with that of their fellow competitors, and to show what they themselves have to offer. In order to understand why certain animals are placed above the others, it is necessary to know what the judge is looking for and how he arrives at his placings.

County fairs and local shows are usually listed in the newspaper. Premium lists, obtainable from the organizers of the show, will list the classes open to exhibitors, as well as the prizes to be won. In addition to classes for purebred sheep there will be classes for market lambs. Sheep may be entered singly or in pens.

Having decided that he has sheep that merit being exhibited at a show, the shepherd must then study the requirements for the breed and the classes in which he is interested. Points are awarded on the basis of a scorecard for each breed and each class, and judging is the business of comparing one sheep with another to see how closely each one compares with the standard for the breed or class.

The judge will know the standard of perfection for the sheep or lambs that he is judging and will inspect each sheep carefully and thoroughly, both by looking at the animal and by feeling it. When he has reached a decision as to which sheep he considers most nearly meets the standard, he will place this sheep first and give his reasons for doing so. A good judge will usually be willing to state these reasons in detail, not only giving the merits of the sheep in order of places but also pointing out the extra good points of other sheep that may place lower because of their average score. He will also point out the weakness of each sheep, and his reasons for awarding lower placings.

If the sheep is to look its best at the show it should not only be well fitted and groomed but should also be trained to handle quietly and to stand still while being judged. The appearance of the handler will also reflect on the sheep to a certain extent: an alert, neatly dressed handler who pays attention to the judge the whole time the class is in the ring will make a better impression than someone who is untidy and whose sheep is out of pose everytime the judge happens to glance at it.

SELECTING SHEEP FOR SHOW

Since it is necessary to feel the sheep to know its condition under all that wool, you first must catch it. Wool is not a handle (pulling on the wool badly bruises the skin), nor are the ears. The sheep from which the prospective show sheep are to be selected should be penned up in a small space, and the ones needed can be caught by grasping the right hind leg up near the flank or under the jaw. A strong walking cane can be used to catch a sheep under the jaw, but never use the small crooks like chicken catchers to grab the sheep round the lower part of the leg.

Before handling the sheep, you should stand back and look over the animal from all angles to get a general impression of its conformation, the make-up of the head, the depth of the brisket, length of legs, top line, bottom line, depth of body, and style.

To make a more detailed examination, first find out the age of the sheep by lifting the lips and looking at the teeth.

You probably know exactly how old the sheep is, but the inspection will accustom it to having its mouth handled. Next study the head to see that the ears are set right and that the head bears the true characteristics of its breed and sex.

To handle the body, lay the hand flat on the animal, feeling with the balls of the fingers, except when it is necessary to grasp any part of the sheep. Starting from the rear of the sheep, stand slightly to the left and rear of the animal and grasp the dock between thumb and forefinger, to note its thickness. A broad dock indicates a meaty backbone. Then feel the top line, checking for straightness of line and thickness of covering. Feel the shoulders, which should be close-set, and the neck, which should be sturdy and not too long. Then find the depth of body, placing one hand on the back just behind the shoulders, and the other on the chest just behind the front legs. Feel the ribs to make sure the chest is deep and that the ribs are well-sprung and meaty. Feel the leg as high up as possible to see how thick it is, and note whether the rump is square or peaked. Finally, place one hand on top of the dock and the other between the two rear

An alert, well-fitted sheep with an equally well turned-out handler.

legs: the distance between two hands gives the depth of the fleshing between the hind legs, or "twist."

The skin of a healthy sheep should be bright and pink and free from bruises. The wool can be inspected for quality, length, and density by parting the fleece on the shoulder and thigh, making the opening vertical and using the flat of the hand. In fine-wool sheep the presence of black fibers is objectionable.

When the qualities of the sheep have been assessed, these can be compared with the standards outlined on the score card. If the animal measures up to the standard required, the next step is to get it "well-fitted," or in good shape for the show.

PREPARING FOR SHOW

A good job of blocking and trimming gives the sheep an attractive appearance and makes a favorable first impression. It shows, too, that the exhibitor thinks enough of his sheep and of the judges' time to do his best. If the sheep are for sale after the show, or if the exhibitor has other sheep for sale, a buyer will have more confidence in sheep that look as though they are healthy and well cared for. The trimming is meant to enhance the good looks of the body underneath, although in some cases the art may be used in an attempt to disguise faults.

A program of good nutrition and careful management is essential to attain good body conformation and fleece quality. Show sheep must be healthy and free from internal and external parasites. Both worms and ticks cause dryness and patchiness of wool, while the anaemia and general loss of condition caused by stomach worms takes all the "bloom" away from the sheep.

Feed must be nourishing and palatable so that the sheep will eat well and be in top condition. To put a high finish on lambs, they should have, in addition to hay, a grain ration of 65 percent oats, 17½ percent wheat bran, and 17½ percent linseed meal; *or* 40 percent corn, 40 percent oats, 10 percent wheat bran, and 10 percent protein supplement. If the roughage consumed is alfalfa, the amount of grain can be reduced by about half.

Yearlings and older sheep will eat about three pounds of grain per head daily and will do well on a ration of 50 percent oats, 40 percent barley, and 10 percent wheat bran. If corn is more easily available, use 40 percent corn, 40 percent oats, 10 percent wheat bran, and 10 percent protein supplement.

Lambs that are to be shown must be pushed from the start and never allowed to lose condition.

If clean, worm-free pastures are available the sheep will get all the roughage they need from grazing, but they will do best if allowed in the barn during the heat of day. In hot weather, grain rations should always be fed early in the morning and in the cool of the evening. Sheep also relish carrots, turnips, and apples. A few apples a day each will be much enjoyed, but beware of a neglected apple orchard with pounds of rotting and fermenting apples on the ground— the sheep will be thoroughly potted after a binge on such goodies.

EQUIPMENT FOR FITTING AND SHOWING SHEEP

The showman's equipment box should contain shears and sharpening stone, grooming brush, scrubbing brush, cards, pocketknife, toe trimmers, bucket, and trimming stand.

The candidate.

Trimming the feet.

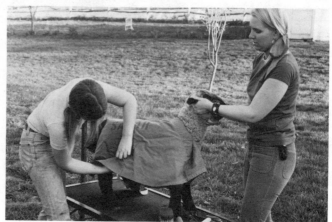

Blanketing.

FEET

Feet are trimmed so that the sheep will stand straight and so that they will be comfortable and graze freely. It is best to trim the feet about every two weeks so that no extensive cutting is necessary and so that there will be no soreness when show time comes.

BLANKETING

Unless the weather is very hot, the sheep should be blanketed to keep dirt and seeds out of the fleece. Blankets can be made of heavy cotton—not burlap—and fitted so that they stay in place but do not constrict the sheep in any way.

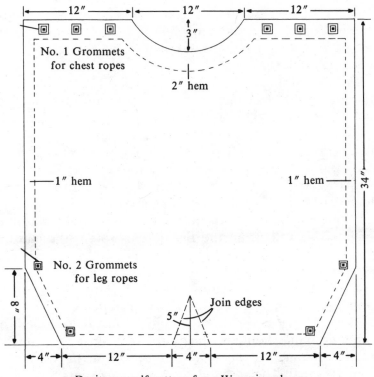

Do-it-yourself pattern for a Wyoming sheep coat. Courtesy of *The Shepherd* magazine.

TRIMMING

This is the most time-consuming part of fitting sheep for show. The trimming must be begun well in advance of show time, and the sheep should require only the smallest touch-ups just before a show. In this way, if a wrong cut is made, there is time for the wool to grow out evenly with the rest of the fleece. The purpose of trimming is to accentuate the most desirable features of conformation and of fleece quality.

The date to start trimming depends on the length of the sheep's fleece. It is usual to start work on mature sheep early in the summer, and then trim them twice before the final touch-up. Certainly the first trim should be more than six weeks before the show. Lambs may not need to be trimmed at all until a few weeks before the show; as much fleece as

189

possible is left on to give the effect of added size. Wethers are trimmed short along the top line to make this area seem firmer, and short trimming also helps keep the animals cool in hot weather.

You should study the animal or animals to be trimmed and have in mind a clear picture of what you want the animal to look like when finished. The first step is to clean the fleece. If possible this should be done by brushing and wiping with a damp cloth, but if the fleece is very dirty the sheep may have to be washed in warm water with 2 tablespoons Cresol (an antiparasitic disinfectant) per gallon. This washing should be done at least two months ahead of show time and should last only long enough to get the dirt out; too thorough a washing removes the yolk from the fleece. This yolk is the natural grease covering the wool fibers and serves to keep the fleece in good condition. Small dung locks can be washed off with a scrub brush and warm water with Cresol. This is better than clipping, which may leave a too-short patch in the fleece.

When the fleece is clean, the animal must be trimmed. A trimming stand should be steady and of a convenient height for the trimmer. If the sheep are skittish or nervous, two straps from the head stand to the rear corners will help prevent them falling off the stand.

First, the fleece is dampened with a wet brush and then thoroughly combed with a curry comb. This combing prepares the surface of the fleece.

The sheep is then trimmed down the back and sides to flatten the back and to give a blocky appearance. This needs careful consideration and a steady hand. The wool on a lamb is trimmed just enough to straighten the top line and to square-off the hindquarters. A more mature sheep can be trimmed down to about ¾ inch along the back and a wether is trimmed to within ½ inch of the hide. Trimming starts at the dock and continues in a straight line to the top of the shoulder. This process is repeated until the whole back is trimmed. Next, the sides are trimmed to give a flatter appearance.

The second trimming, a few weeks later, is a more thorough job. First the fleece is thoroughly dampened, and

Trimming.

Currying.

then brushed and carded to straighten the fibers and pull out loose ends. Overly brisk carding will spoil the fleece. The brush should be dipped frequently in Cresol solution. After the brushing, carding separates the fibers and straightens them so that they can be trimmed evenly. To use the sheep card properly, place it in the fleece so that when it is lifted up with a revolving motion the fibers are separated and combed out parallel to each other. The sides, rump, and under line should be carded to give an appearance of full development.

Next, the sheep shears are used to trim the fleece to an even length. This process may have to be repeated several times, brushing, carding, and trimming until the fleece is smooth and even.

When the appearance of the body is satisfactory, it is time to give some attention to the head. In some breeds, wool must be trimmed from around the eyes and ears, and wool on the sides of the face may need tidying. Some wool trimmed off the throat will make the front end look wide and deep.

191

The final trimming consists of more brushing and carding, along with final snips with the shears to ensure a smooth, even fleece.

At the show, the fleece is dampened slightly and very lightly carded to give a fresher appearance. A touch of specially formulated black dressing is applied to the black wool on sheep that have black faces and feet, such as Hampshires and Suffolks. Finally, a very light dressing of oil is rubbed into the fleece to give the wool extra luster.

Trimming to give a blocky appearance.

THE SHOW

In preparing to take sheep to a show, remember that health certificates are always required for sheep that are to be shipped out of state. The certificate must accompany the bill of lading if the sheep are shipped by public carrier, or must accompany the sheep if transported by owner or handler, and must be signed by a licensed veterinarian.

The sheep should arrive at the show at least two or three days before the day of exhibition. Use clean straw for bedding, and allow plenty of room in the pens for each animal to bed down in comfort. Provide the necessary partitions for penning ewes and rams separately, and for keeping separate rams that are not used to being with each other.

Take with you enough feed for the whole trip, and feed the sheep very lightly immediately before and during traveling. Treat the animals quietly, and do not park the transporting vehicle in hot sunlight or where the sheep would be exposed to cold winds.

As soon as the animals reach the show site, they should be unloaded and placed in the pens with clean bedding. Exercise them daily, perhaps while the pens are being cleaned.

Have a sign for the pen giving the breed of the sheep and the name of the exhibitor.

By the time the day of the show comes, the sheep should be well accustomed to being handled and to being led. It will be used to standing still squarely on all four feet and not moving away when a stranger approaches.

The exhibitor should complete the final touching up well enough in advance so that when the time comes to enter the ring, he is prompt and has a chance to get a good position in the ring.

Carding.

Trimming the face.

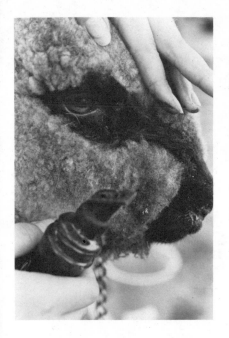

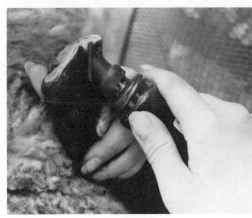

Applying black dressing.

The exhibitor stands to the left front of the animal, holding the sheep by grasping the wool lightly under the chin, leaving the right hand free for keeping the sheep posed properly. A push against the animal's breast with the knee will keep the sheep standing firmly with the back straight.

When the judge walks in front, the exhibitor should step aside to give the judge a clear view; when the judge moves around, the exhibitor changes his position to be out of the way, but still keeping the sheep properly set up.

A good showman remains unruffled, is equally courteous whether a winner or a loser, and is considerate of the other exhibitors.

Upon return home, keep the sheep isolated from the remainder of the flock for three weeks in order to be sure that the show sheep have not brought any disease or parasites home with them.

While the sheep are on the show circuit they should be kept in peak condition all the time, but when the shows are over they must be gradually let down by providing plenty of exercise and increasing the amount of bulk in the ration in relation to the amount of grain.

9

Home
Butchering

There are many advantages to being able to butcher one's own lambs and sheep. It is said that meat butchered at home always tastes better, and there may be a good foundation for this. When an animal goes to the slaughterhouse, it is driven, loaded, penned, and worried in unfamiliar surroundings; fright and exhaustion make it tense and nervous and have the effect of toughening the meat. When the animal is butchered at home under proper conditions, however, there is no tension—it remains happy and relaxed.

Lambs or sheep to be butchered should be penned up overnight with fresh water, but no feed; food takes approximately five days to pass through the whole digestive tract, so there is no danger of acute hunger or shrinkage if feed is withheld, and the job of removing the gut will go much more smoothly if the stomach is not too full. Never butcher an animal that has received any kind of medication within the last two weeks. For instance, if the animal bloated and had received bloat medicine, this would contaminate the carcase; if it were being treated with penicillin for foot rot, this would have the same effect. Sheep that have been wormed or treated for external parasites may not be slaughtered for human consumption within a prescribed period, depending on which medicine was used. When cutting up the carcase, it is important not to puncture the stomach or gallbladder, as their juices will give the meat a bad taste that is hard to wash off. Try to do the butchering on a cool day and work as quickly as possible, because gases soon form in the stomach.

Having penned up the sheep, it is time to assemble all the necessary tools and to make sure they are clean, that the knives are sharpened, and that such things as pulleys, if used, are in working order.

You will need the following equipment:

- Large pail of water for washing equipment.
- Meat saw.
- Sharp butchering knives and steel to sharpen them while working.
- Bowls for catching the blood: one for the gut and one for the liver and kidneys.
- Butchering "cradle" or sturdy table.
- Some means of hanging the carcase.
- Large tub or sink for washing the carcase.

Liver and gallbladder. Lamb liver is as tender and tasty as the best calves' liver, and equally nutritious. The gallbladder, indicated by the point of the knife, must be removed carefully so as not to puncture the organ and thereby taint the meat.

Butchering tools.

• A roll of paper towels for wiping hands and working surfaces.

THE PROCEDURE

When all the equipment is assembled, lead the sheep as near as possible to the butchering cradle by putting the left hand under the throat and the right hand under the opposite flank or at the dock. Handle gently and do not pull the wool, because rough handling causes bruising and will spoil the meat. If the sheep is used to grain or to being hand-fed, it can be tempted to walk along by following a helper with an apple or a bucket of grain.

Shoot the sheep in the center of the forehead with a .22 rifle, and immediately lift the body onto the cradle. With a

Correct way to hold a sheep.

A.

B.

C.

D.

sharp, pointed knife, grasp the head of the sheep and insert the point of the knife in the throat, cutting from within to the outside (see A). After bleeding, the lamb is placed on its back. The skin is cut down the center of the belly, cutting only just through the skin, to the point of the jaw. Remove the head at the first joint (B). If the head is to be used, it should be skinned before removing from the body and washed in cold water. The brains and tongue should be taken out and put in a separate pan, covered with cold water.

At the breastbone, the incision divides (C), and the pelt over the breast is pulled backward. It may be necessary to loosen the skin with the knife. The neck should be skinned as far as possible. The forelegs are skinned to a point just above the knees (D), and the lower part of the leg is removed. With the neck, breast, and foreshanks skinned, the pelt over the front part of the belly is removed by making a fist with one hand and pressing against the flesh, while holding the skin with the other hand (E). The skin is then slit down the center, and the breastbone cut with the saw (F).

Next, the throat is split down the center and the windpipe and gullet removed (G).

The hind legs are then skinned, and the back hooves are removed at the hoof head to preserve the tendons of the hind legs for hanging the carcase (H and I).

The pelt over the rear of the belly is fisted off and the skin cut down the center. An incision is made in the hind legs between the tendons and, if no meat hooks are available, strong cords can be threaded through. By now the under side of the sheep is skinned; the head, gullet, and front feet are removed; and the carcase is ready to be hung.

The pelt is then fisted off the sides of the carcase, after which the fist should be forced in over the top of the shoulder (J). It is important to remove the pelt downward from the shoulder and upward from the hind leg to avoid tearing the "fell," or inner skin.

The pelt is forced upward over the loin (the skin may have to be loosened around the tail with a knife) and the skin is then pulled downward and off the carcase (K and L).

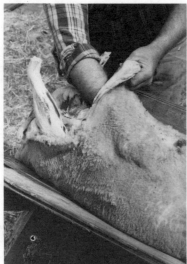

E.

F.

G.

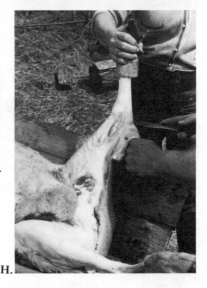

H.

I.

J.

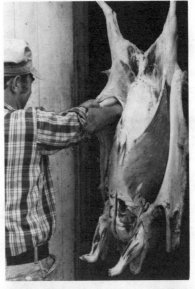

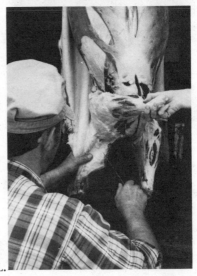

K.

L.

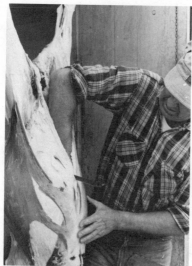

M.

N.

O.

P.

To remove the intestines, an incision is made through the belly wall, and the fist is forced into this, holding the knife (M). The forearm forces the gut away from the knife as the belly wall is cut downward. The viscera are then removed, leaving the kidney in the carcase. The paunch and liver are removed, the paunch is allowed to drop into a pail, and the liver is placed in cold salted water. The diaphragm is cut through and the heart and lungs removed. The heart goes in the salted water, and the lungs will probably go in with the dog meat unless you have plans for making a Scotch haggis.

The carcase is then washed in tepid water (N), making sure the insides are clean.

The pelt should be well salted with ordinary table salt (O) and hung over a fence to dry until it can be tanned.

The carcase should then be covered with netting to keep flies off and hung in a cool place for at least twenty-four hours. Ten days at 36°F. is good.

CUTTING LAMB AND MUTTON

Cut several one-inch-thick slices from the neck. Saw across the backbone and remove the two legs, separate the legs, and remove the shanks (P and Q).

Remove the sirloin; this may be left whole for a small roast, or separated for chops. (R).

Separate the breast pieces with a saw and remove the rack from the shoulders. This can be left whole for rack of lamb or separated for chops. Saw through the backbone to separate the shoulders.

Slice the liver about ¾ inch thick. Lamb liver is very tender and rich in iron. Remove the kidneys from their fatty covering. Wash the heart well to make sure all blood is removed, and leave whole.

Lamb shoulders may be left whole to make a square, rather bony roast that is a trial to cut, or it may be boned and either rolled or left with the pockets to be stuffed. Shoulders from small lambs that would make rather small roasts may be

Q.

R.

cut into chops one inch thick. The breast of lamb makes excellent stew; it may be cut into small pieces, or the meat may be lifted from the bone with a sharp knife to make a pocket for stuffing.

Ribs can be separated for lamb riblets for barbecuing. The loin may be made into chops, or left whole for rack of lamb. The loin is not difficult to bone: the meat is completely removed from the backbone with a sharp knife and then

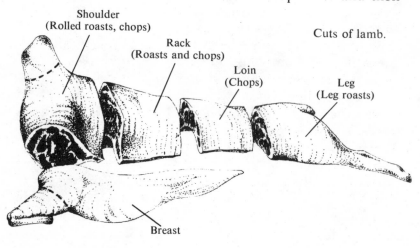

Shoulder
(Rolled roasts, chops)

Rack
(Roasts and chops)

Loin
(Chops)

Leg
(Leg roasts)

Cuts of lamb.

Breast

rolled lengthwise and tied at two-inch intervals. This small roast may be stuffed or cut into boneless lamb chops.

Cut the ribs into chops, or leave them whole for rack of lamb. The legs, if left whole, make a roast fit for a king. The bone may be left in, or the meat may be slit up the leg bone and the bones removed and the roast tied.

THE EASTER LAMB MARKET

In the urban centers and in the Northeast there is a strong market at Easter time for milk-fed lambs. These usually weigh about forty pounds live weight and are sold straight from the mother, before weaning and at premium prices. The peak demand comes immediately before Easter for the Italian market and for Passover lambs for the Jewish market, and again at the Greek Orthodox Easter for the Greek traditional meals. Such lambs are usually sold directly from the farm to the consumer, or, where a large supply is available, to a dealer who specializes in the Easter lamb trade.

Lambs for the Passover market must be slaughtered according to Jewish dietary rules. The killing is performed by a rabbi according to a prescribed method, and because kosher meat may not be held longer than 72 hours, slaughter must take place near the area where they are to be consumed. Because of this, lambs for the kosher market are usually shipped live to markets near the consuming centers.

The skins from these small lambs are ideal for making toys or pillows. About eight skins would be needed to make a man's sheepskin jacket.

COOKING WHAT YOU'VE GROWN

Mutton is usually defined as a sheep of more than one year of age; it is not, as some believe, an aged animal too old for further breeding or an old ram that has no more interest in his ewes. There is another way to define mutton. In lambs the break-joint, or joint just above the ankles of the forefeet, can be severed. In yearlings, this joint becomes more porous and dry, and in mature sheep the cartilage will no longer break, making it necessary to take off the foot at the ankle. This makes a round (or spool) joint, and any animal having a spool joint is classed as mutton.

Lamb and mutton should both be cooked at low temperatures; it is hot burning fat that gives the meat its strong smell, and a roasting temperature of no more than 325°F. is recommended. A sprinkling of lemon juice over a lamb or mutton roast gives it an excellent flavor, but heavy seasoning with garlic or spices overwhelms the delicate flavor of the meat.

In addition to being a flavorful and delicious food, lamb contains high-quality proteins, minerals, and vitamins, particularly the B group of vitamins. Lamb and mutton, too, are the only meats acceptable to all castes and religions, and are frequently recommended by the medical profession as being the most digestible of all meats.

WHAT MAKES LAMB CHOPS SO EXPENSIVE?

A hundred-pound lamb on the hoof produces about forty-two pounds of retail cuts, and of these cuts, about 3¾ pounds are lamb chops. How much does it cost to produce these chops? The following estimates must of necessity be very approximate.

Keeping one bred ewe from breeding time to weaning of the lamb or twin lambs.	Hay	10 bales	$10.00
	Grain	50 lb	5.00
	Straw	1 bale	1.00
Vaccine (T.A.T. or *Clostridium perfingens*)			.25
Creep feed for lamb(s) 150 percent is a fair average crop, or 1½ lambs per ewe.	Grain	60 lb	5.00
	Hay	½ bale	.50
			$21.75

This total is just for feed and routine veterinary supplies and does not allow for lighting, heat lamps, heat tapes, salt, routine maintenance of equipment, labor, or office time (record keeping, bill paying, tax work, etc.).

If an average ewe produces 1½ lambs and these are sold at 100 pounds live weight, she has produced 150 pounds of lamb which may possibly sell at 40 cents per pound (more or less depending on the market), so for his time and expenditure the producer has made $60.00 for his lambs from one ewe, less the $21.75 shown above, less the other costs listed.

If the buyer pays $40.00 for a 100 pound-lamb that is to produce 42 pounds of retail cuts, he must sell this meat for 95 cents per pound before he begins to cover his cutting and handling costs. Some of the cuts of lamb that are least in demand, such as neck and breast, must be sold for a lower price, thus making it necessary to sell the popular cuts of chops and legs for top prices.

LEG OF LAMB

This is the most popular lamb roast. It weighs from 4 to 8 pounds with the bone in; a hanging-weight carcass of 35 pounds will yield about a 5-pound leg of lamb. Larger legs of lamb are usually divided into leg and two steaks, taken from the loin end of the roast. These steaks can be broiled. The roast should be cooked at 325°F. for about thirty minutes per pound and is done when the meat is just faintly pink. For best results, leave the fell, or thin papery skin, on the roast while it is cooking. If the meat is very fat, trim this off and sprinkle liberally with lemon juice. Mixing ¼ teaspoon of basil with lemon juice will give the meat a pleasant, tangy flavor. Other herbs that blend well with lamb are rosemary, mint, dill, and tarragon.

In England, red currant jelly is served with lamb. This has a slightly milder taste than the mint jelly more commonly served in the United States and can also be used to make a glaze by combining ½ cup of jelly, ½ cup of dry sherry, ½ cup of ketchup, and a pinch of rosemary or basil, heated together until the jelly melts. The glaze is brushed over the lamb during the last hour of cooking time.

SHOULDER OF LAMB

A lamb shoulder is a difficult roast to carve, and for the sake of the carver's nerves we recommend boning this roast. This is done by inserting a sharp knife along the blade bone and lifting the meat away from both sides of the bone, then inserting the knife along the "arm" bone and cutting this loose. The pockets in the roast may then be stuffed with any meat stuffing, such as ground lamb and pine nuts, ground lamb and rice, ground veal, sausage, or prepared poultry stuffing. The pockets are then secured with skewers or lacing pins and the meat roasted at 325°F. for about thirty to thirty-five minutes per pound (until tender). The shoulder may also be rolled and tied, with or without any of the above stuffings.

SIRLOIN OF LAMB

This small roast may be treated in the same way as lamb shoulder.

LAMB CHOPS

Tender, tasty lamb chops should be cut between one and two inches thick. Either broil until tender, or brown in hot fat in a frying pan, place on a rack in a roasting pan, and then cook in the oven until tender.

Milanese Lamb Chops

8 lamb chops	1 cup rice
1 cup chicken broth	2 tbsp butter
½ cup tomato sauce	1 tbsp Parmesan cheese
1 tsp chopped parsley	salad oil

Cook the rice in the combined chicken broth, tomato sauce, and butter (with salt and pepper to season) until the rice is tender. Brush the chops with salad oil and brown them for five minutes on each side, or until tender. Season with salt and pepper. Add the Parmesan cheese to the rice mixture and serve on a platter surrounded by the lamb chops. Garnish with parsley.

LAMB NECK SLICES

This cut is delicious braised in chicken stock and served with tiny peas, new potatoes, and young carrots. Most recipes calling for lamb shoulder chops are equally good for lamb neck slices.

LAMB SHOULDER CHOPS
Skillet Lamb Chop Dinner

4 shoulder lamb chops	¼ tsp marjoram
salt and pepper	1 chicken bouillon cube
2 medium onions	4 potatoes, peeled and
1 box (10 oz) frozen peas and carrots	quartered

Brown the chops on both sides in large skillet, sprinkle with salt and pepper, and remove. To the drippings in the skillet, add marjoram, bouillon cubes, 1 tsp salt, ⅛ tsp pepper, and ½ cup cold water. Bring to a boil, stirring. Add onions, potatoes, peas and carrots, and seasoning; cover, and cook seven minutes. (Courtesy, *The Shepherd*.)

Raisin Rice Lamb Chops

6 lamb shoulder chops	3 tsp cooking oil
1 tsp salt, dash of pepper	¼ cup water
¾ cup uncooked rice	½ cup raisins
6 lemon slices, ¼" thick	¼ cup chili sauce
dash of rubbed or ground sage	

Brown the lamb chops in fat, pour off drippings, and season with salt and pepper. Add water, cover tightly, and cook slowly for thirty minutes. Add the raisins to the rice and cook the rice according to directions on the package. Place a lemon slice on each lamb chop and cook slowly fifteen minutes longer, or until done. Serve on rice. (Courtesy, *The Shepherd.*)

LAMB SHANKS

These are tender when braised, oven cooked, or used for casseroles.

Apricot Lamb Shanks

4 lamb shanks	¼ cup light brown sugar
2 tbsp salad oil	1 tsp salt
1 12-ounce can beer	1 tsp ground ginger
1 cup dried apricots	¼ tsp pepper

Brown the shanks in oil in a large skillet or Dutch oven. Drain off the drippings. Place the lamb shanks in a large baking dish. Combine the remaining ingredients and pour over the shanks. Cover and bake at 350°F. for two hours or until meat is tender. (Courtesy, *The Shepherd*.)

Swedish Lamb Shanks

4 lamb shanks, or more	1 tbsp grated horseradish
2 tbsp butter or margarine	2 tsp grated orange rind
1 tsp paprika	1 tsp rosemary, crumbled
1 cup sliced onions	1 tsp salt
1 cup sliced mushrooms	¼ tsp pepper
½ cup dry vermouth or white wine	¼ cup chopped fresh dill
½ cup sour cream	

Melt butter or margarine in a large saucepan. Sauté three or four shanks at once until golden brown. Pour off accumulated fat, and return lamb to saucepan. Sprinkle with paprika. Add onions, mushrooms, wine, horseradish, orange rind, rosemary, salt, and pepper. Cover and cook over medium-low heat for one hour and fifteen minutes. Remove the shanks to a warm serving platter. Add sour cream and dill to the drippings in the saucepan, and heat to serving temperature over low heat, stirring constantly. Do not allow to boil. Pour over the lamb and serve.

BREAST OF LAMB

This is an economical dish. If the meat is inclined to be fatty, a stew made with breast of lamb is better if cooked and allowed to cool overnight. The fat may then be removed and the stew reheated the next day.

Stuffed Breast of Lamb

With a sharp boning knife, make a pocket in the lamb breast and stuff with vegetables or any of the stuffings given for shoulder of lamb. Cook until the meat is tender.

Irish Stew

3 lb stewing lamb	4 allspice berries
1 large onion	2 tbsp chopped parsley
3 large carrots, sliced	1 cup cubed turnips
salt and pepper	4 lb potatoes

Place meat in heavy electric frying pan, add the allspice berries and 2 cups water, then simmer, covered, for two hours. Add vegetables, season with salt and pepper, and use more water if necessary. Cook until the vegetables are done. The gravy may be thickened if desired. This dish is good served with dumplings.

Lancashire Hot Pot

2 lb lamb neck, lamb stew	salt and pepper
or cut-up lamb shoulder	2 lb potatoes
3 kidneys	1 tbsp butter
1 cup chicken stock	1 large onion

Grease an ovenproof baking dish, and put in a deep layer of sliced potatoes. Arrange the lamb on top of the potatoes, and then slice the kidneys and place them on top of the meat. Season with salt and pepper. Add the rest of the potatoes, arranging the top layer in overlapping slices. Brush with butter and bake two hours in a moderate oven. Remove the cover after the first hour of baking. Serve in the dish in which it was baked.

Lamb Stew

Stews, ragouts, and casseroles are great dishes for using the less tender cuts of meat and provide an opportunity for the creative cook to go to work. Slow cookers make wonderful stews, and almost any lamb stew recipe adapts well to this kind of cooking; vegetables and meat will be tender and tasty. If a recipe calls for sour cream, add it just before the dish is served. Lamb shoulder, cut off the bone and cubed, is especially good in stews. The following is a basic stew recipe from the Lamb Council kitchens.

1 ¼ lb lamb shoulder	2 tbsp salad oil
⅔ cup sliced onion	2 tsp salt
¼ tsp coarse ground black pepper	¾ tsp ground allspice
¼ tsp ground ginger	1 can (8 oz) tomato sauce
1 tbsp flour	
1 can (1 lb) peas and carrots, undrained	

Heat the oil in a large skillet or saucepan, then brown the lamb on all sides. Add the onion and brown lightly. Drain off the fat. Stir in salt, the spices, and ½ cup water, and simmer covered for about two hours. If necessary, add more water during the simmering period. Blend 2 tbsp water into the flour. Add this flour mixture, tomato sauce, and the peas and carrots to the lamb. Cook until the stew is thickened (about fifteen minutes). Serve with boiled potatoes, if desired.

LAMB SPARERIBS AND RIBLETS

The breast of lamb is the sparerib; when these ribs are cut apart between the bones, the result is riblets. These ribs are tender and tasty and very economical. They can be broiled, barbecued, braised, and stewed. Cooking time will be a few minutes to brown the meat and then one to 1½ hours cooking in liquid. For barbecuing, ribs may be partly cooked in the oven first.

Lamb Riblets with Vegetables

3 lb breast of lamb, cut into riblets	2 tbsp fat
	⅛ tsp pepper
2 tsp salt	6 small white onions
1½ cups water	1 lb green beans
8 whole carrots	¼ cup vinegar

Melt the fat in a Dutch oven, add the lamb, and brown on all sides. Add salt, pepper, and water, then cover and simmer forty-five minutes. Add the vegetables, cook thirty minutes. Add the vinegar and cook for a further thirty minutes.

GROUND LAMB AND MUTTON

These can be used in any meat loaf recipe; they can also be used in combination with ground beef or veal.

Lamb Loaf

2 lb ground lamb	½ cup milk
1 cup cracker crumbs	1½ tsp salt
½ cup chopped onions	dash of pepper
¼ cup chopped green pepper	¼ tsp garlic salt
	¼ tsp marjoram
¼ cup ketchup	1 cup grated chedder cheese
2 eggs, beaten	

Pour the milk over the cracker crumbs. Add lamb, salt, pepper, garlic salt, marjoram, onions, green pepper, ketchup, and eggs. Mix thoroughly. Pack in a 9 x 5-inch loaf pan. Bake at 325°F. for 1½ hours. Combine the crumbled bacon and grated cheese, and sprinkle over the meat mixture. Continue baking for fifteen minutes or until the cheese is melted.

Busy Day Casserole

Try making this one ahead of time and serving it at the end of shearing day.

2 lb ground lamb	2 medium onions,
1 can mixed vegetables,	chopped
undrained	1 cup ketchup
2 cups mashed potatoes	several slices cooked bacon

Brown the meat and onion, pour off the fat, and add vegetables and ketchup. Top with mashed potatoes, place slices of cooked bacon on top, and cook at low heat for about thirty minutes.

Shepherd's Pie

This recipe is taken from an old English cookbook. Shepherd's pie may also be made with ground meat and sometimes with vegetables added.

½ lb cold mutton (or lamb)	1 lb mashed potatoes
1 tbsp butter	½ pint gravy, or stock
1 tsp parboiled and	salt and pepper
finely chopped onion	

Cut the meat in small thin slices. Melt half the butter in a frying pan, add the potato, salt, and pepper, and stir over medium heat until thoroughly mixed. Grease a baking dish, line the bottom thinly with potatoes, put in a layer of meat, then a layer of potatoes, and meat, finishing with potatoes. Season each layer lightly with salt and pepper. Pour in the stock or gravy. Dot the top of the pie with the remaining butter. For a change, slices of tomato and bacon can be arranged on top of the pie. Bake at 325°F. until the surface is well browned, about thirty minutes.

Sheeps' Tongues

From the same book comes this method of cooking sheeps' tongues.

4 sheeps' (or lambs') tongues	2 slices bacon, cooked
stalk of celery	1 carrot
1 onion	1 tbsp butter
½ turnip	6 peppercorns
bouquet garni (parsley, thyme, bay leaf)	
1 cup stock	

Soak the tongues in saltwater for two hours. Then put the tongues in a pan of fresh water and bring to a boil, drain, and brown. Slice the vegetables, put them into a saucepan with butter, bouquet garni, and peppercorns, lay the tongues on top, then cover and cook gently for twenty minutes. Add hot stock to nearly cover the vegetables, lay the bacon on top of the tongues, cover, and cook gently for 2½ hours or until tongues are tender. Discard the vegetables.

When the tongues are cooked, skin them. They may then be served on hot spinach, with mashed or fried potatoes.

LAMB HEART

Remove the tubes and wash thoroughly. Cut the clean heart in pieces, place in cold water, and bring gently to boil. Simmer, covered, until tender, about two hours.

Lamb Heart with Vegetables

2 hearts	1 onion, chopped fine
stalk celery, diced	salt and pepper
1 tbsp butter	1 tbsp flour

Cook the hearts as above. When the meat is tender, add vegetables and salt and pepper to taste. Cook for thirty minutes. With 2 tbsp of the liquid, mix the flour and butter into a paste and add this to the heart and vegetables. Bring gently to a boil, stirring constantly. Serve on toast, mashed potatoes, or rice.

LAMB KIDNEYS

Remove fat and membranes, and soak one hour in cold saltwater. Drain dry. Slice ¼ inch thick, sprinkle with salt and pepper, and dredge in flour. Fry the kidneys in butter or margarine for a few minutes, turning frequently. Serve with bacon.

"PAUNCH AND PLUCK"
Haggis

This is the traditional national dish of Scotland and is usually eaten on November 30th, the feast day of Saint Andrew, the patron saint of Scotland. At such a feast, the dish would be borne aloft on a platter into the banquet hall, escorted by pipers, making a tour of the whole room.

1 sheep's or lamb's paunch and pluck (the pluck is the liver, heart, and lungs)	1 cup oatmeal
	2 tbsp salt
	½ nutmeg, finely grated
1 lb finely chopped beef suet	juice of 1 lemon
1 lb finely chopped Spanish onions	1 tsp pepper
1½ pints of stock or good gravy	

Soak the pauch for several hours in saltwater, then turn it inside out and wash it thoroughly in several waters. Wash the pluck, cover the liver with cold water, boil it for 1½ hours, and at the end of ¾ hour add the heart and lights (kidneys and sweetbreads). Chop half the liver coarsely, then chop the rest of the liver and lights finely, and mix all together. Add the oatmeal, suet, onions, salt, pepper, nutmeg, lemon juice, and stock. Fill the paunch with these ingredients and sew up the opening, taking care that sufficient space is left for the oatmeal to swell: if the paunch is overfull it may burst. Put the haggis into boiling water and cook gently for about three hours; during the first hour it should be occasionally pricked with a needle to let the steam escape. To serve, slit the haggis crosswise at the top and pour some whisky over it, allowing some to soak into the meat. Set light to the whisky. Serve each diner a shot glass of whisky, to be drunk along with the first taste of the dish. From then, you are on your own.

Index

INDEX